福建菜

家傳滋味

Fujian's Flavour

下個月將是母親93歲生日，
我把這本書送她作為生日禮物，
小小心意，不能答謝她疼愛我的萬份之一，
唯盼上天保佑母親更快樂、更健康！

～茵茵

Valerie Ong

王陳茵茵

「在烹調中思考、創新、分享──是我最心滿意足的事！」

"In the cooking process, I think, create and share – it is the most fulfilling experience I've ever had."

福建泉州人，兒時隨家人來港定居，年青時為生活拼搏，捨棄了她的烹飪興趣，嚐盡人生甘、苦、澀，在此人生三味中，奮鬥出一番事業。

現在，她從工作退下來，全身投入人生的第四味──享受入廚、旅遊、跳舞，與至親好友共聚的美味人生。平日，王太最愛在家烹調美食，每當在外嚐到好吃的菜式，不管是福建家鄉菜、中菜、西菜、東南亞菜或蛋糕甜點，都會喜孜孜地回家鑽研做法，並配搭不同食材加以創新，弄出特色佳餚，與摯愛分享。

和藹親切的王太，每逢假日，必定親手炮製美食，與丈夫、兒媳們、孫兒女一聚天倫，在嘻哈談笑、逗孫為樂中，享受人生第五味──快樂、圓滿。

Born in Quanzhou in Fujian province, China, Valerie came to settle in Hong Kong with her family when she was a child. In the early years, she worked hard for a living and was forced to give up culinary art, her favourite pastime. As she savoured the sweetness, bitterness and hardship in life, her career took off.

Now that she has stepped down from the front line of her job, Valerie got to indulge herself in the fourth taste of her life – the leisurely time in the kitchen, travelling, dancing, and sharing quality time with her family and close friends. Valerie enjoys cooking at home most. Whenever she tastes any good food in restaurants, no matter it's Fujian home-style food, Chinese, Western, Southeast Asia, cakes or even desserts, she would go home and experiment in the kitchen, so as to replicate the same good taste. She never shies away from being creative by tossing in different ingredients for her very own variations and sharing them with her loved ones.

Valerie is such a gentle and amiable person. She insists on spending time with her husband, children and grandchildren every Sunday, when she'd make them heavenly good food. In the laughter and harmony, she watches her grandchildren grow up and this is the fifth taste of her life – fullness and happiness.

Preface

自 序

福建省位於中國東南沿海，不論是山珍和海味都非常豐富。自唐宋以來，泉州已開放對外通商，由各地傳入烹飪技巧，至今演變成口味精細，注重調味，善於滷、燜、煮，尤善於清鮮海味的製作，因地域關係，很多閩菜也是台菜的根源。

我的家鄉是福建泉州。解放那年，隨家人來到香港，雖然年紀小，但兒時的回憶還很清晰，尤其家中的食物常令人念念不忘。

記得冬天飯桌上常有火鍋，是那種用炭燒的銅火鍋，裏面滾着煮熟的食物。偶爾爸會把烏魚籽切片，掃上黃酒，在炭爐上架了鐵絲網，烤得可香！弟妹眾多，如能分得一片也就心滿意足了。飯後在火爐上烤着柑，一陣陣果皮香，聽着母親的琴音，享受的是那滿滿的溫馨吧！

夏天把西瓜放進後院的井裏，鎮得冰涼，還有荔枝，我們幾個孩子吃荔枝，要比賽，把外皮剝去，留着白膜不破的才叫高手，我常贏得獎，可多吃幾個。

Located on the southeastern coast of China, Fujian province is home to both terrestrial and marine delicacies. Ever since Tang and Song dynasties, Quanzhou in Fujian province has been opened to international trade. Being a hotchpotch where different culinary cultures meet, Fujianese cuisine is the perfect distillation of the numerous cooking methods to become the delicate taste as it is today. Fujianese cooking emphasizes on seasoning, with a resourceful repertoire of marinating, braising and simmering recipes. As a province by the sea, freshest seafood is also a common item on the menu. Many Taiwanese dishes are variations of Fujianese food.

I was born in Quanzhou, Fujian. In 1949, when the Communist Party won the civil war, I left my hometown with my family to settle in Hong Kong. Although I was young then, I still remember my childhood vividly, particularly those episodes related to food.

I remember the winters when the family ate out of a coal-burning copper steamboat, in which food was boiled vigorously. Sometimes dad would grill sliced dried mullet roe on charcoal stoves. He would brush some Shaoxing wine on the mullet roe, place it on a wire mesh and put it over the charcoal. That smell was so divine. Since I had many siblings, I considered it my lucky day if I was able to get one whole slice. After dinner, the smell of mullet roe would be replaced by the smell of mandarin oranges on the grill. The sweet tangy smell and mom's piano made the perfect memory of homely warmth.

I remember the summers when we chilled watermelons in the well in our backyard. I remember lychee peeling competitions between us kids as we strove to keep the white pith intact while peeling. I often won, getting to savour several more lychees than my siblings.

大門外小販在賣芋頭，挑在擔裏的是切開兩半的芋頭，蒸得冒煙，三姆買給我吃時，小販灑上鹽花，香極了！祖母穿的上衣裏縫了個大口袋，裝着零錢，拉着我的手上街買油條，還有説不出名堂的炸物。

最興奮是給人做「花童」，以及去戲院登台表演歌舞，完了去「吃桌」，吃些甚麼已不太記得，但肉刈包是我記憶中最美味的，吃不完，還可以帶些回家。

我們的院子和大學中間是一片龍眼樹林。有一天看到解放軍在摘龍眼，叫：「勇哥、勇哥，丟些給我們吃吧！」肥大的龍眼就從天而降了。龍眼是我家鄉盛產的水果。

那次隨父母到小霞美看望外祖母，我和二妹是坐在竹籃裏挑着走的，給轉得頭暈眼花，可是辛苦也吵着要去，因為外祖母家的果園種了很多花果，那大手掌般的棕巴梨又甜又多汁。外祖母人很本事，做得一手好菜，還有糕點，只有那鼠麴粿是我不愛的，那名字「鼠」，多可怕！而且黑黑的不好看，怎樣都不吃。回想起來，真不懂事呀！怎麼就不會體貼老人的心意？那次也是最後一次看到外祖母，因為過了

I remember the street foods. Just outside our front door sat a hawker who carried steamed taro halves with his shoulder pole. My aunt often bought them for me. Before serving it, the hawker would sprinkle salt on them. I can still vividly remember that smell. I remember my grandmother always wore shirts with a big pocket in the front jiggled with spare change. She would hold my hand and take me to street vendors and get deep fried dough sticks among other anonymous fried goodies.

My fondest memories involved being a flower girl at a wedding and performing on stage. It was because dinner parties often followed these occasions. Although I don't quite remember what exactly was served at these dinner parties, I do remember distinctly those innumerable Chinese buns stuffed with meat. Even after stuffing my face with many, many of them, there would still be many more left over for me to bring home.

I remember the forest of longan trees that sat between our backyard and the university. One day, we saw some soldiers picking the longans and we yelled "Brave soldier, brave soldier, throw us some!" And down from the sky rained plump, succulent longans. Longan is a famous fruit of my hometown.

I remember the visits to my grandmother in Xiaoxiamei. My sister and I each sat in one of the two bamboo baskets hung from either end of a shoulder pole and were carried to my grandmother's. It was a bumpy ride, but despite our motion sickness, we were still adamant about going because there were many fruit trees in my grandmother's orchard. I especially loved those trees where juicy sweet pears as big as an adult's palm hung from. Moreover, my grandmother was an amazing cook and I loved everything she made, except for this Cudweed Cake. I wonder why anyone would eat anything made from weed and it was pitch black and frighteningly ugly. There was no way I was touching it, let alone eating it. In retrospect, I regret it. It was rude of me to reject my grandmother's good intentions.

一個禮拜，我們便離開家鄉來了香港。

來港後的日子過得很忙亂，上小學二年級到中學，轉了無數學校，終於能如願在我心儀的中學領了畢業証書。放下書包，結婚、生子、做事，忙忙碌碌，不知時日飛逝，欣慰的是兒子們事業有成，有了美滿家庭。

現在過的是退休的清閒日子，有時做些家鄉菜，大人一桌，小孩一桌，都回家來吃得熱鬧。孫兒們一句：「嫲嫲，好好味呀！」讓我樂上半天。好友們聚聚吃吃聊聊天，是人生樂事！家鄉的菜餚大多是母親和三姆煮的家常菜，覺得傳統的味道是值得保留和流傳的。

家人好友們，特別是小妹的多方鼓勵，以及丈夫無限的支持，促使我能出版這本食譜，我衷心感謝！還有家中的幫傭阿 May 是我的得力助手，應記一功。

It was also the last time I saw her; for a week later, we left for Hong Kong.

Life became busy and chaotic after arriving in Hong Kong. From primary 2 to secondary school, I transferred frequently between schools, finally graduating from the secondary school I wanted. After graduation came marriage, children and work; busy and hectic, time somehow waltzed right on by me. Now, I can only take pride and joy at the fact that my sons have a successful career and families of their own.

Nowadays I enjoy the life of a retiree, spending most of my time cooking for my family and friends. Simply hearing my grandchildren saying "Grandma, this is so yummy!" or being able to cook for my old friends makes me happier than ever. The dishes you find in this cookbook are mostly family recipes passed down from my mother and aunt, as I believe traditional taste is worth preserving most.

Lastly, I'd like to thank my family and friends for helping me write this cookbook.. I'd especially like to thank my younger sister for her encouragement and my husband for his unwavering support. And finally, to my helper May, for being my assistant in the kitchen, thank you!

2010. 12月

自序的英文版是長孫女王皓怡暑假回港時幫忙翻譯，謝謝！
This English preface was translated by my granddaughter Jennifer Ong when she spent her summer vacation in Hong Kong. I express the most heartfelt gratitude to her.

目錄 CONTENTS

福建 • 主菜
Fujian Entrees

福建 • 湯羹
Fujian Soups

福建 • 清粥小菜
Fujian Congees and Stir-Fried

福建 • 小吃麵飯
Fujian Snacks and Staples

福建 • 甜品
Fujian Desserts

重量單位換算表
Conversion Guide

1 兩（tael）	=	38 克（g）
4 兩（tael）	=	150 克（g）
8 兩（tael）（半斤）	=	300 克（g）
1 斤（catty）	=	600 克（g）
1 磅（lb.）	=	450 克（g）
1 安士（ounce）	=	28 克（g）
1 杯（cup）	=	250 毫升（ml）

福建 • 主菜
FUJIAN ENTREES

回到家，
細嘗閩南鄉土風味主食，
暖胃透心，回味綿長！

Come home to an authentic Minnan dinner
with authentic rustic entrees
that warm the stomach and the heart

泉州潤餅宴

Quanzhou
Wrap Feast

有幸得蔡瀾先生讓我轉刊他有關薄餅的大作，特別在此向他再三說句：「多謝您」。

「潤餅來自春餅，到了春天各地都有拜祭春神獻上的春盤，春盤即五辛盤，包括大蔥、小蒜、韭菜、苦菜和香菜五種生辛。有開五臟、去伏氣之功效，讓人體小宇宙的陰陽和大宇宙互相感應，借立春後長出來的野菜新芽來淨化身體。

春盤逐漸發展為春餅，用五辛配合其他食材，包上一層薄餅來吃，就成為閩南人的習俗。台灣和閩南分隔不開，發揚光大後也分台南和台北式，與本來的泉州薄餅不同，又與福建人來到南洋包的相異。

薄餅的餡，台灣人主要用高麗菜（椰菜），而南洋人則改為沙葛，馬來人叫的 Mankung。原則上，是愈炒愈入味，加上各種配料的準備，做起薄餅來總得花上兩三天，不是特別慶典是不做的。

泉州薄餅，最具代表性。泉州當年為世界最大的海港，為海上絲綢之路的出發點，到了元代，更與歐亞非一百多個國家有海上貿易，阿拉伯人、印度人、中東人、歐洲人，都成為泉州的居民。

所以泉州的薄餅，從基本的五辛之外，加上紅蘿蔔的維他命、荷蘭豆的葉綠素、生蠔的鈣與鋅、豆乾的蛋白、滸苔的鉀、香菜的治高血壓和花生的營養。」

蔡瀾

這是薄餅，泉州話叫「潤餅」，吃時各人自己動手包，特別熱鬧，親切有氣氛，乃喜慶、過節、歡聚的佳餚。吃過的親友們常一吃難忘，邀我再做。席間，單吃潤餅一道菜已能盡歡，如再加番薯粥及幾碟下粥小菜就更滿足了。

下列是我家常用的食材，因材料多是蔬菜、海鮮，不油膩，故一家老少皆宜吃。材料準備需時，最好提前一天預備，把材料分開炒，各自調味，調味不必重手，因為全部煮熟的材料混合後可試味，有需要才加以調整。泉州潤餅鹹甜適中，不煮得太糊，必須乾身，香氣四溢才是佳品，吃完盤底不剩油不剩汁才算合格。

如餡料吃不完也不浪費，次日把餅皮蒸軟，包了炸，即是春卷；或簡單地稍加熱水加蓋翻熱透，可配粥、麵、飯享用。

這道菜，看似複雜，其實並不難做，成功率高，擺上桌後很討人歡喜，是值得推薦的家宴特色菜。我希望不論是否福建同鄉，大家都來一試！

【材料】

1. 紅蘿蔔
2. 冬筍
3. 高麗菜（椰菜）
4. 荷蘭豆
5. 蒜仔
6. 韭菜
7. 唐芹
8. 芫茜梗
9. 香菇
10. 木耳
11. 豆乾
12. 蝦仁
13. 蟹肉煎蛋
14. 魚肉
15. 瑤柱
16. 滸苔
17. 花生粉
18. 春卷皮

* 各項材料預備及煮熟後，分開盛於碟內，先不要摻雜一起，方便最後處理。

【配料】

芥辣、辣椒醬、蒜白、芫茜葉、滸苔、花生粉、蒜仔掃 4 條

【 Ingredients 】

1. carrot
2. bamboo shoots
3. white cabbage
4. snow peas
5. Chinese leeks
6. chives
7. Chinese celery
8. coriander stems
9. dried black mushrooms
10. wood ear fungus
11. dried beancurd
12. shelled shrimps
13. crabmeat scrambled egg
14. de-boned fish meat
15. dried scallops
16. green hair algae
17. ground peanuts
18. spring roll wrappers

* After preparing and cooking the ingredients, keep them separately on plates. Do not mix them together.

【 Condiments 】

mustard, chilli sauce, white parts of Chinese leeks, coriander leaves, green hair algae, ground peanuts, 4 stems of white parts of Chinese leeks for spreading the sauces

紅蘿蔔
Carrot

【材料】

紅蘿蔔 2 公斤
乾葱頭 4 湯匙（切片）
糖 3 湯匙
生抽 1/4 杯
油 1 杯
雞湯 1 杯（後下）

【 Ingredients 】

2 kg carrot
4 tbsps sliced shallot
3 tbsps sugar
1/4 cup light soy sauce
1 cup oil
1 cup chicken stock (added at last)

【做法】

1. 紅蘿蔔洗淨，去皮，刨幼絲。

2. 燒熱油，下乾葱頭用中火爆香，下紅蘿蔔絲炒軟，加調味料及雞湯炒勻，加蓋煮至熟透（期間開蓋拌炒，以防燒焦）。

【 Method 】

1. Rinse and peel the carrot. Then shred it finely.

2. Heat some oil in a wok. Stir fry the shallot over medium heat until fragrant. Add shredded carrot and stir fry until soft. Add seasoning and chicken stock. Stir well. Cover the lid and cook until done. Stir once in a while to prevent it from burning. Set aside.

【小秘訣 • TIPS】

● 紅蘿蔔的份量比其他蔬菜較多，因紅蘿蔔的甜度高，是整道菜的主角。

● 炒紅蘿蔔時，建議多下油炒，因紅蘿蔔頗吸油。

● 炒煮時，下糖增加甜度，以生抽代替鹽更能增加鮮味。

● 加入雞湯加蓋燜煮，可縮短烹調時間，也令味道更佳。

● The portion of carrot is bigger than other vegetables. As it has natural sugar in it, carrot is the key ingredients for the wraps.

● When you stir fry carrot, you may use a bit more oil than usual. It's because carrot picks up much oil.

● The sugar is there to accentuate the sweetness of carrot. To balance it off, we need some salty seasoning. I used soy sauce instead of salt for an extra caramel-like flavour.

● Cooking the carrot in chicken stock with the lid covered helps shorten the cooking time. The chicken stock also adds another dimension of flavour.

冬筍
Bamboo shoots

【材料】

冬筍 4 個（或罐裝竹筍 2 罐，每罐 650 克）
油 4 湯匙
雞粉、鹽各 1 茶匙
紹酒、糖各 1 湯匙

【Ingredients】

4 fresh winter bamboo shoots (or 2 cans of bamboo shoots, weighing 650 g each)
4 tbsps oil
1 tsp chicken bouillon powder
1 tsp salt
1 tbsp Shaoxing wine
1 tbsp sugar

【做法】

1. 冬筍切去頭部較粗部分，棄去，去掉外殼，整個放入滾水焓 15 分鐘，取出，放在水喉下沖至冷。

2. 冬筍切片（別太薄，以免影響爽脆度），再切絲。

3. 燒熱油，下冬筍炒香，放入調味料，濳酒炒勻即可。

【Method】

1. Cut off the rounder end of each bamboo shoot. Peel the outer leaves and use just the tender heart. Blanch the whole bamboo shoots in boiling water for 15 minutes. Drain. Rinse under a cold tap until cold.

2. Slice the bamboo shoots. (Do not slice them too thin or they won't be as crunchy.) Then cut into strips.

3. Heat some oil and stir fry the bamboo shoots until fragrant. Add seasoning. Sizzle with wine. Set aside.

【小秘訣 • TIPS】

- 挑選冬筍時，可選手掌般大小，形如半彎形，外殼沒有裂開的為佳。

- 若買不到新鮮冬筍或春筍，可用罐裝代替，省略飛水步驟即可。

- Pick bamboo shoots in a crescent shape about the size of your palm. The best ones should not have any cracks on the surface.

- If you can't get fresh winter or spring bamboo shoots in the market, use canned ones instead. Just skip the blanching step.

【小秘訣 • TIPS】

- 椰菜選外型呈扁身、色澤淡綠的較佳。

- Pick white cabbages that are flat in shape and light green in colour.

高麗菜（椰菜）
White cabbage

【材料】
高麗菜 2 個（約 1.5 公斤）
乾葱頭 4 湯匙（切片）
油半杯

【調味料】
生抽 1/4 杯
糖 1 湯匙

【 Ingredients 】
2 white cabbages (about 1.5 kg)
4 tbsps sliced shallot
1/2 cup oil

【 Seasoning 】
1/4 cup light soy sauce
1 tbsp sugar

【做法】
1. 椰菜去心，洗淨，抹乾水分，切幼絲。

2. 燒熱油，下乾葱頭爆香，下椰菜絲炒至軟，放入調味料拌勻炒熟，隔去汁液即可。

【 Method 】
1. Cut open the cabbages. Cut off the hard stems. Rinse and wipe dry. Finely shred them.

2. Heat oil in a wok. Stir fry shallot until fragrant. Add shredded cabbages and stir fry until soft. Add seasoning and stir well. Cook until cabbage is done. Drain off the liquid. Set aside.

荷蘭豆
Snow peas

【材料】
荷蘭豆 800 克
油 2 湯匙
乾葱頭 2 湯匙（切片）

【調味料】
雞粉半湯匙
生抽 2 湯匙
紹酒 3 湯匙
瑤柱汁或清雞湯 1 杯

【Ingredients】
800 g snow peas
2 tbsps oil
2 tbsps sliced shallot

【Seasoning】
1/2 tbsp chicken bouillon powder
2 tbsps light soy sauce
3 tbsps Shaoxing wine
1 cup broth from steamed dried scallops or chicken stock

【做法】
1. 荷蘭豆洗淨，撕去硬筋，切幼絲。

2. 燒熱油，爆香乾葱頭，下荷蘭豆炒香，最後加入調味料炒熟，盛起。

【Method】
1. Rinse the snow peas. Tear off the tough veins. Shred finely.

2. Heat oil in a wok and stir fry shallot until fragrant. Put in snow peas and stir well. Add seasoning and stir until done. Set aside.

【小秘訣 • TIPS】
- 炒荷蘭豆時潛入酒，可辟除草青味。

- Adding Shaoxing wine to snow peas helps remove the grassy taste.

蒜仔
Chinese leeks

【材料】

蒜仔（蒜苗）1.6 公斤
油半杯
鹽 1 茶匙

* 預留 4 條蒜白作為掃醬料使用，長約 4 吋，頭部剺成十字紋。
* 預留 4 杯生蒜白絲，備用（吃前浸於冰水）。

【做法】

1. 蒜仔洗淨，瀝乾水分，切去頭部硬筋及尾段，棄去。蒜白部分拍扁，切粒備用。
2. 鑊內燒熱油，放入蒜仔炒至軟身，灑入鹽拌勻後，盛起。

【Ingredients】

1.6 kg Chinese leeks
1/2 cup oil
1 tsp salt

* set aside 4 stems Chinese leeks (white part only) for spreading the sauces. They should be about 4 inches in length. Make a crisscross cut at one end.
* Set aside 4 cups of shredded white part of the Chinese leeks to be served raw. Keep them in iced water before serving.

【Method】

1. Rinse the leeks. Drain well and cut off the root end. Tear off the tough veins. Pat the white parts flat and dice them.
2. Heat oil in a wok. Stir fry the leeks until soft. Sprinkle with salt. Serve.

韭菜
Chives

【材料】
韭菜 800 克
油 3 湯匙
糖 1 湯匙
鹽 1 茶匙

【Ingredients】
800 g chives
3 tbsps oil
1 tbsp sugar
1 tsp salt

【做法】
韭菜洗淨，瀝乾水分，切小段，用油、鹽及糖略炒軟（別炒太久，以免質感粗韌）。

【Method】
Rinse the chives. Drain well. Cut into short sections. Stir fry the chives in oil with sugar and salt until soft. Do not overcook them, or they will get tough.

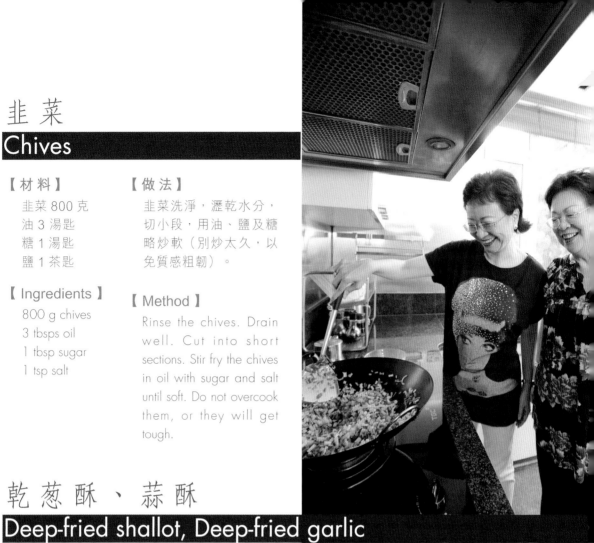

乾葱酥、蒜酥
Deep-fried shallot, Deep-fried garlic

【做法】
1. 乾葱頭及蒜頭切片，用油炸香，取出，瀝乾油分，分別放於廚房紙上，吸去餘油，待涼。
2. 享用前放入玻璃瓶儲存，吃時取出，保持香酥。

【Method】
1. Slice the garlic and shallot. Deep fry in oil until golden. Drain. Leave them on paper towel to absorb the oil until cool.
2. Before serving, store the deep-fried shallot and garlic in an air-tight container to keep them crispy.

【小秘訣 • TIPS】
- 炸乾葱頭及蒜頭時，見葱頭或蒜頭呈硬身及微黃色，立即取出，置於漏勺瀝乾油分。
- 乾葱酥及蒜酥盛起後，顏色會持續加深，若色澤太焦黃會帶苦澀味。

- In the deep frying process, take the garlic and shallot out from the oil when they turn slightly golden and stiff. Drain the oil in a strainer ladle.

- After they are lifted off the oil, the remaining heat will keep on cooking them and they will keep getting darker in colour. If you overcook them, they would taste bitter.

唐芹
Chinese celery

【材料】
唐芹 500 克
油 2 湯匙
鹽 1 茶匙

【Ingredients】
500 g Chinese celery
2 tbsps oil
1 tsp salt

【做法】
唐芹洗淨，瀝乾水分，切粒，用油及鹽炒香，盛起。

【Method】
Rinse the Chinese celery. Drain well. Dice it and stir fry in oil with salt. Set aside.

芫茜梗
Coriander stems

【材料】
芫茜 400 克

【Ingredients】
400 g coriander

【做法】
芫茜梗切成小段（芫茜葉保留，備用），毋須炒煮，最後只需與其他材料拌勻即可。

【Method】
Cut the coriander stems into short lengths. Set aside the leaves for later use. The coriander stems don't need to be cooked. Just stir it in with other ingredients later.

香菇
Dried black mushrooms

【材料】
香菇 10 朵

【調味料】
生抽 2 湯匙
麻油 2 茶匙
糖 1 湯匙

【 Ingredients 】
10 dried black mushrooms

【 Seasoning 】
2 tbsps light soy sauce
2 tsps sesame oil
1 tbsp sugar

【做法】
1. 香菇用水浸透，去硬蒂，擠乾水分，下調味料拌勻，蒸 30 分鐘至香菇入味。

2. 香菇切片後，再切成幼絲，盛起備用。

【 Method 】
1. Soak the black mushrooms in water until soft. Cut off the stems. Squeeze dry. Add seasoning and stir well. Steam for 30 minutes until the mushrooms are flavourful.

2. Slice the mushrooms. Then finely shred them. Set aside.

木耳
Wood ear fungus

【材料】

木耳 80 克
鹽及糖各 1 茶匙
油 3 湯匙
水 2 湯匙

【Ingredients】

80 g wood ear fungus
1 tsp salt
1 tsp sugar
3 tbsps oil
2 tbsps water

【做法】

1.　木耳用水浸透，剪去硬蒂，切細絲。

2.　燒熱油，下木耳絲炒香，最後加入調味
　　料拌勻，盛起備用。

【Method】

1.　Soak the wood ear fungus in water until
soft. Cut off the hard stems. Finely shred it.

2.　Heat oil in a wok. Stir fry shredded wood
ear fungus until fragrant. Add seasoning
and stir well. Serve.

豆乾
Dried beancurd

【材料】
豆乾（大塊）6 件
乾葱頭 3 湯匙 （切片）
油半杯

【調味料】
五香粉 1 茶匙
鹽 1 茶匙
紹酒 3 湯匙
糖 2 湯匙
生抽 2 湯匙

【Ingredients】
6 large pieces dried beancurd
3 tbsps sliced shallot
1/2 cup oil

【Seasoning】
1 tsp five-spice powder
1 tsp salt
3 tbsps Shaoxing wine
2 tbsps sugar
2 tbsps light soy sauce

【做法】
1. 豆乾洗淨，抹乾水分，切片後再切成絲。
2. 燒熱油，爆香乾葱頭，下豆乾炒勻及乾透，加入調味料炒香，盛起。

【Method】
1. Rinse the dried beancurd. Wipe dry. Slice them and then shred them.
2. Heat oil in a wok. Stir fry shallot until fragrant. Add dried beancurd. Stir fry until done. Add seasoning. Stir until fragrant. Set aside.

【小秘訣 • TIPS】
- 必須傾入多些油炒豆乾，因豆乾頗吸油。
- 炒豆乾時若見滲出水分，必須去掉水分，才容易將豆乾炒至乾身。
- Use more oil than usual when you stir fry the dried beancurd. It tends to pick up much oil.
- In case the dried beancurd gives water in the stir frying process, make sure you drain the liquid well. It's hard to just cook off all the water. The dried beancurd should be fairly dry after being fried.

游水鮮蝦
Shelled shrimps

【材料】
　游水鮮蝦 600 克
　薑 2 片

【 Ingredients 】
　600 g live shrimps
　2 slices ginger

【做法】
1.　燒滾水，放入薑及鮮蝦焓熟，取出。
2.　鮮蝦去殼、去腸，切成小丁，備用。

【 Method 】
1.　Boil the water. Put in 2 slices of ginger. Blanch the shrimps until done. Drain.
2.　Shell the shrimps and devein them. Dice finely. Set aside.

蟹肉煎蛋
Crabmeat scrambled egg

【材料】
大花蟹 2 隻（約 1 公斤）
蛋 6 隻（拂勻）
乾葱頭 3 湯匙（切片）
油 3 湯匙

【蟹肉調味料】
胡椒粉及鹽各半茶匙
薑汁 1 茶匙
紹酒 1 湯匙

【蛋汁調味料】
生抽 1 湯匙

【做法】
1. 花蟹蒸熟，拆肉，加入蟹肉調味料拌勻。

2. 蟹肉、乾葱頭、蛋汁及調味料拌勻，傾入熱鑊炒香，盛起備用。

【 Ingredients 】
2 large swimmer crabs (about 1 kg)
6 eggs (whisked)
3 tbsps sliced shallot
3 tbsps oil

【 Seasoning for the crabmeat 】
1/2 tsp ground white pepper
1/2 tsp salt
1 tsp ginger juice
1 tbsp Shaoxing wine

【 Seasoning for the eggs 】
1 tbsp light soy sauce

【 Method 】
1. Steam the crabs until done. Shell them and pick the crabmeat free of shells. Add the seasoning for the crabmeat. Mix well.

2. In a mixing bowl, mix crabmeat, shallot, whisked eggs and seasoning. Fry in a wok until set and lightly browned. Set aside.

鮮魚肉
De-boned fish meat

【 材料 】

大鱲魚或馬頭魚 2 條（約 1.4 公斤）
薑 5 片
生抽及紹酒各 1 湯匙
油 5 湯匙

【 醃料 】

薑汁 1 湯匙
紹酒 1 湯匙
鹽 2 茶匙
胡椒粉 1 茶匙

【 Ingredients 】

2 large snappers or tilefishes (about 1.4 kg)
5 slices ginger
1 tbsp light soy sauce
1 tbsp Shaoxing wine
5 tbsps oil

【 Marinade 】

1 tbsp ginger juice
1 tbsp Shaoxing wine
2 tsps salt
1 tsp ground white pepper

【 做法 】

1. 鮮魚洗淨，抹乾水分，於魚身剁兩刀，下醃料抹勻在魚身及內腹待 2 小時。

2. 燒熱油 3 湯匙及薑 3 片，下魚煎至兩面金黃色，盛起，拆肉。（緊記完全去掉魚骨，拆肉後用手檢查清楚，確保沒有小骨。）

3. 燒熱油 2 湯匙及薑 2 片，加入魚肉、生抽及紹酒炒香，盛起備用。

【 Method 】

1. Rinse the fishes. Wipe dry and make a couple slashes on the fleshiest part of the fish. Add marinade and rub well on both the insides and outsides of the fishes. Leave them for 2 hours.

2. Heat 3 tbsps of oil in a wok. Put in 3 slices of ginger. Fry the fishes until both sides golden. Set aside to let cool. De-bone the fishes. Make sure you pick the fish meat free of any bone. Double check with your fingers to remove any small bone.

3. Heat 2 tbsps of oil in a wok. Put in 2 slices of ginger. Stir fry the fish meat in light soy sauce and Shaoxing wine until fragrant. Set aside.

乾 瑤 柱
Dried scallops

【 材料 】
瑤柱碎 2 杯
清雞湯 3 杯

【 Ingredients 】
2 cups dried scallops (not necessarily in whole)
3 cups chicken stock

【 做法 】
1. 瑤柱碎用清雞湯浸 40 分鐘，連清雞湯一
   併蒸 1 小時。

2. 去掉瑤柱邊的硬枕，撕成幼絲備用。瑤柱
   汁保留，可作炒荷蘭豆或紅蘿蔔之用。

【 Method 】
1. Soak the dried scallops in the chicken stock
   for 40 minutes. Steam the dried scallops in
   the chicken stock for 1 hour.

2. Remove the tough tendons on the dried
   scallops. Tear the dried scallops into fine
   shreds. Set aside the steaming broth for stir
   frying snow peas or carrot.

滸苔
Green hair algae

【材料】
滸苔 120 克
油 2 杯
砂糖 250 克

【 Ingredients 】
120 g fresh green hair algae
2 cups oil
250 g sugar

【做法】
1. 調至最小火,燒熱油半杯,趁溫油下 1/4 份量滸苔,快速用鑊鏟拌炒及撥至鬆散,滸苔短時間會吸油變得香酥(緊記別炒至焦黃,帶苦味)。

2. 關火,盛起,趁熱與糖拌勻,備用。

3. 餘下的滸苔重複以上步驟,分 4 次炒香。

【 Method 】
1. Heat 1/2 cup of oil in a wok over the lowest heat. Add 1/4 of the algae when the oil is warm. Quickly stir it with a spatula to scatter it. The algae will pick up the oil and turn crispy instantly. Do not overcook it. It tastes bitter when it's browned.

2. Turn off the heat. Set aside the algae. Stir in sugar while still hot. Set aside.

3. Repeat step 1 and 2 four times until all algae is cooked.

【小秘訣 • TIPS】
- 滸苔於北角春秧街有售。包潤餅品嘗,滸苔是必需的材料,增加美味之餘,還可保餅皮不破爛。

- 建議以上滸苔的份量分 4 次炒香,較易操作。

- Green hair algae are available from Chun Yeung Street market in North Point, Hong Kong Island. To make authentic Quanzhou wraps, green hair algae are a must-have ingredient. Not only do they add flavour, they also keep the wrapper from breaking.

- For the amount listed here, I suggest stir frying the algae in 4 separate batches. They can be handled more easily this way.

花生粉
Ground peanuts

【材料】
　花生 1.2 公斤
　砂糖 250 克

【Ingredients】
　1.2 kg peanuts
　250 g sugar

【做法】
1. 花生連膜放入白鑊內炒至香脆（或購買市面出售的連殼炒花生），待涼，去外膜。
2. 花生分數次放入攪拌機內，攪拌成幼粉狀，下砂糖徹底拌勻，備用。

【Method】
1. Fry shelled peanuts in a dry wok. Leave them to cool. Rub off their red coats. (Or you may get those fried peanuts in shells. Then just shell them and rub off their red coats.)
2. Blend the peanuts in a blender in several batches until fine. Add sugar and mix well. Set aside.

【小秘訣 • TIPS】
- 緊記別用油炸花生，因為攪拌花生時容易釋出油分，難以磨成幼粉，粗粒狀的花生欠美觀，而且香味不足。

- Do not deep fry the peanuts in oil. The peanuts carry much oil themselves. If you deep fry them, they'd pick up even more oil when you blend them. The peanuts would be lumpy and cannot be ground finely. Not only is the presentation affected, but also the aroma of the dish.

春卷皮
Spring roll wrappers

【小秘訣 • TIPS】

- 春卷皮可預早一天在麵店訂購（銅鑼灣鵝頸街市坤記麵店有售），指明是大張薄餅皮，一斤約 22 張。
- 北角春秧街也有現售的春卷皮，但數量不多。
- 買回來的春卷皮是一整叠的，享用前預先一張張撕開來，然後用保鮮紙包好及蓋上濕毛巾，春卷皮才不容易乾裂。

- Order spring roll wrappers one day ahead from noodles shop (such as Kun Kee Noodles in Bowrington Road Market, Causeway Bay). Specify you need large sheets of spring roll wrappers with about 22 sheets in 600 g.

- Some stalls in Chun Yeung Street Market, North Point also carry spring roll wrappers in limited quantities.

- To prepare the spring roll wrappers, separate each wrapper from the stack. Then stack them together again. Wrap them in cling film and cover them in a damp towel. The wrappers won't dry out and crack this way.

芥辣醬、辣椒醬、蒜白、
芫茜葉、蒜仔掃

Mustard,
chilli sauce,
white parts of Chinese leeks,
coriander leaves,
Chinese leeks for spreading
the sauces on the wraps

【做法】

1. 芥辣醬及辣椒醬盛於碟上，吃前先用保鮮紙包好。

2. 蒜白及芫茜葉切碎。

3. 於蒜仔的蒜白部分剝成十字狀，作為塗抹醬料之用，蒜仔掃共 4 條，方便各人使用。

【Method】

1. Put mustard and chilli sauce separately on plates. Wrap them in cling film before serving.

2. Chop the white parts of Chinese leeks and coriander leaves.

3. Four stems of white parts of the Chinese leeks, about 4 inches long each, have been reserved for spreading the sauces. Make a crisscross cut on one end of each. Place them around the table.

潤餅菜綜合做法
Filling Assembly

【小秘訣 • TIPS】

- 各項材料分別炒煮妥當後，由於份量較大（可供 24 人享用），建議將材料各取一半拌炒，方便烹調。

- This recipe here makes 24 servings which could be a bit difficult to fit into a household wok. To make your life easier, you may put in half of each ingredient in a wok at one time, doing the assembly in two batches.

【做法】

將第 1 至 15 項材料（見第 16 頁）傾入鑊內，不停翻炒至熱透及均勻，香氣撲鼻時，灑入麻油、胡椒粉、乾葱酥及蒜酥，以增加香氣，試味至鹹甜適合自己口味，盛於大碗上桌分享。

【Method】

Put ingredients 1 to 15 (refer to p.16). Keep stirring until fragrant, well mixed and heated through. Sprinkle with sesame oil, ground white pepper, deep fried shallot and garlic. Taste it and season accordingly. Transfer into a large bowl and serve.

餐桌上的擺設
Table setting

1. 大碗盛起已炒至香氣撲鼻的潤餅菜。

2. 碗內放上已拌糖的花生粉。

3. 碗內放上拌糖的滸苔。

4. 生蒜及芫茜各1碗。

5. 辣椒醬配上蒜仔掃。

6. 芥辣醬配上蒜仔掃。

7. 碟上放好已撕開一張張的春卷皮，摺成三角形狀。

8. 每位客人預備大碟一隻，方便包捲潤餅，配上濕紙巾，並奉上福建名茶鐵觀音。

1. Place the large bowl of stir fried filling on the table.

2. Add ground peanuts with sugar stirred in.

3. Add green hair algae with sugar stirred in.

4. Arrange 1 bowl of raw Chinese leeks and 1 bowl of coriander leaves.

5. Place a plate of chilli sauce on the table and put a stem of white parts of Chinese leeks for spreading.

6. Place a plate of mustard on the table and put a stem of white parts of Chinese leeks for spreading.

7. Fold each spring roll wrapper into a wedge shape. Arrange them on a plate.

8. Set one large flat plate for each guest for rolling the wraps. Provide wet wipes for your guests. Serve with quality Iron Buddha tea from Fujian province.

潤餅包捲法
Wrap Assembly

【小秘訣 • TIPS】

- 盡量多放潤餅菜及其他材料，通常每人可吃 3 至 6 卷，吃至飽足。
- 配合潤餅宴，可預備清淡的湯水或稀粥小菜伴吃；甜品可配花生豆花湯。

- Try to put in as much filling and condiments as you can in each wrap. Prepare 3 to 6 wraps for each guest.

- The wraps are the main course on their own. To complete the meal, you may serve a light-tasting soup or a light congee. For dessert, try peanut sweet soup with soft beancurd dessert.

【做法】

春卷皮一張鋪平，在餅皮中央橫向先放一層薄薄的滸苔，灑上花生粉，再放上潤餅菜，灑上芫茜及蒜白，另塗抹醬料在餅皮上，包捲成筒形，用雙手拿着，細意品嘗。

【Method】

Lay a sheet of spring roll wrapper flat on a plate. Arrange a thin layer of green hair algae across the diameter of the wrapper at the centre. Sprinkle with ground peanuts. Put some filling on. Sprinkle with coriander and white parts of Chinese leeks. Spread your favourite sauces on the wrapper. Roll it up into a barrel shape. Serve.

酒燜雞 Braised Chicken
in Shaoxing Wine

金針菜亦稱為「忘憂草」，在福建菜裏廣被採用，有減壓療效，亦有傳古時的肺病者吃金針菜至痊癒，也說能滋潤皮膚、止痕癢。

【材料】

雞 1 隻
栗子 10 粒
金針菜半杯
雲耳（已浸發）2 杯
葱 4 條（切段）
蒜頭、乾葱頭各 6 粒（拍扁）
薑 2 片
油 2 湯匙

【調味料】

鹽及胡椒粉各 1 茶匙
蠔油、生抽及老抽各 3 湯匙
冰糖 2 湯匙
雞粉 1 茶匙
紹酒 1 杯（後下）

【 Ingredients 】

1 chicken
10 chestnuts
1/2 cup dried day lily flowers
2 cups cloud ear fungus (re-hydrated)
4 sprigs spring onion (cut into short lengths)
6 cloves garlic (gently crushed)
6 cloves shallot (gently crushed)
2 slices ginger
2 tbsps oil

【 Seasoning 】

1 tsp salt
1 tsp ground white pepper
3 tbsps oyster sauce
3 tbsps light soy sauce
3 tbsps dark soy sauce
2 tbsps rock sugar
1 tsp chicken bouillon powder
1 cup Shaoxing wine (added at last)

【做法】

1. 雞洗淨，斬成小塊。栗子去殼，放入滾水內煮 15 分鐘，取出，趁熱去皮。

2. 金針菜浸水，剪去硬蒂，打成結。雲耳浸水，去掉硬蒂，切大塊。

3. 燒熱油下乾葱頭、蒜頭、薑及葱段爆香，放入雞及栗子拌炒，加調味料再炒香，轉入已預熱煲仔內，潷入紹酒煮滾，改小火燜煮 30 分鐘，最後加入金針菜及雲耳拌勻再煮 15 分鐘，原煲上桌。

【Method】

1. Rinse the chicken and chop into small pieces. Set aside. Shell the chestnuts and blanch in boiling water for 15 minutes. Peel while still hot.

2. Soak the day lily flowers in water. Cut off the hard ends. Tie each flower into a knot. Set aside. Soak the cloud ear fungus in water until soft. Cut off the tough stems. Cut into large chunks.

3. Heat oil in a wok and add shallot, garlic, ginger and spring onion. Stir fry until fragrant. Put in the chicken and chestnuts. Stir well. Add seasoning (except the wine) and cook further. Transfer into a preheated clay pot. Sizzle with Shaoxing wine. Bring to the boil and turn to low heat. Simmer for 30 minutes. Add day lily flowers and cloud ear. Stir well and cook for 15 more minutes. Serve in the clay pot.

【小秘訣 • TIPS】

- 燜煮時用小火，肉質會鮮嫩；若用高火快煮，肉質容易變硬。

- 金針菜打結後，吃起來有脆度，而且又美觀。

- For braised dish, always cook the meat slowly over low heat so that it remains succulent and tender. If you cook it over high heat quickly, the meat turns tough.

- Tying the day lily flowers into knots gives them a lovely crunch. They also look better.

紅蟳翡翠

Crabmeat
and Angled Loofah
on Rice Vermicelli

【 材 料 】
　米粉 100 克
　小膏蟹 1 隻
　絲瓜 200 克
　蛋白 1 隻
　蒜頭 1 粒（切茸）
　乾葱頭 2 粒（拍扁）
　油適量

【 煮 米 粉 料 】
　水 4 杯

【 絲 瓜 調 味 料 】
　鹽及糖各半茶匙
　紹酒 1 湯匙
　水 1/4 杯

【 蟹 肉 調 味 料 】
　鹽及胡椒粉各 1/4 茶匙
　紹酒 1 湯匙

【 芡 汁 】
　清雞湯 2 杯
　水 1 杯
　生粉 1 茶匙

【 Ingredients 】
　100 g rice vermicelli
　1 small female mud crab
　200 g angled loofah
　1 egg white
　1 clove garlic (grated)
　2 cloves shallot (gently crushed)
　oil

【 For cooking rice vermicelli 】
　4 cups water

【 Seasoning for angled loofah 】
　1/2 tsp salt
　1/2 tsp sugar
　1 tbsp Shaoxing wine
　1/4 cup water

【 Seasoning for crabmeat 】
　1/4 tsp salt
　1/4 tsp ground white pepper
　1 tbsp Shaoxing wine

【 Thickening glaze 】
　2 cups chicken stock
　1 cup water
　1 tsp caltrop starch

【做法】

1. 絲瓜洗淨，去皮及去籽，切塊。蛋白打勻，備用。

2. 膏蟹洗淨，隔水蒸 20 分鐘，取出，拆肉，加入調味料拌勻。

3. 米粉剪成段，放入滾水中煮熟，取出，瀝乾水分，置於深碟內。

4. 燒熱油 2 湯匙爆香乾葱頭，下絲瓜及調味料炒熟，取出，排放米粉上。

5. 燒熱油鑊，爆香蒜茸，下芡汁煮滾，放入蟹肉再煮滾，拌入蛋白，澆在絲瓜上即成可口鮮美的佳餚。

【Method】

1. Rinse the angled loofah. Peel and seed it. Cut into slices. Whisk the egg white and set aside.

2. Rinse the crab. Steam for 20 minutes. Leave it to cool and shell it. Pick the crabmeat free of any shell. Add seasoning. Stir well.

3. Cut the rice vermicelli into short lengths. Blanch in boiling water until done. Drain. Place on a deep dish.

4. Heat 2 tbsps of oil in a wok. Stir fry shallot until fragrant. Add angled loofah and seasoning. Stir fry until done. Arrange over the bed of rice vermicelli in the deep dish.

5. Heat the wok and add oil. Stir fry garlic until fragrant. Pour in the thickening glaze. Bring to the boil. Add crabmeat and bring to the boil again. Stir in the egg white and cook until half set. Dribble over the angled loofah. Serve.

【小秘訣 • TIPS】

- 選用絲瓜，因喜愛它的清甜爽口（可用其他瓜菜代替）。

- 以米粉鋪於碟底，可吸收蟹汁鮮味，蟹膏的紅配絲瓜的翠，故名為「紅蟳翡翠」，是一款具備色香味的佳餚。

- I used angled loofah for this dish because of its crunchiness and refreshing taste. You may use other gourd or vegetables of your choice instead.

- I put rice vermicelli on the bottom of the dish to drench them in the crab sauce and flavour. It is truly a dish that tastes, looks and smells equally great.

紅蟳糯米飯

Glutinous Rice with
Mud Crab

上桌品嘗，給人喜氣洋洋的感覺，單吃糯米飯已很美味，加上陣陣蟹香，蟹汁沁入飯內，令糯米飯更軟糯可口。

【材料】
　紅蟳（膏蟹）2 隻
　糯米 3 杯（量米杯）
　水 2 杯（量米杯）
　小蝦米 2 湯匙
　臘腸 2 條（切粒）
　香菇粒 3 湯匙
　乾蔥頭 4 粒（切片）
　蔥 1 條（切粒）
　薑 4 薄片
　油 2 湯匙

【調味料】
　生抽及糖各半湯匙
　紹酒、蠔油及老抽各 1 湯匙
　胡椒粉 1 茶匙
　清雞湯半杯

【 Ingredients 】
　2 female mud crabs
　3 cups glutinous rice (measured with the cup that comes with your rice cooker)
　2 cups water (measured with the cup that comes with your rice cooker)
　2 tbsps dried small shrimps
　2 Chinese preserved sausages (diced)
　3 tbsps diced dried black mushrooms
　4 cloves shallot (sliced)
　1 sprig spring onion (diced)
　4 thin slices of ginger
　2 tbsps oil

【 Seasoning 】
　1/2 tbsp light soy sauce
　1/2 tbsp sugar
　1 tbsp Shaoxing wine
　1 tbsp oyster sauce
　1 tbsp dark soy sauce
　1 tsp ground white pepper
　1/2 cup chicken stock

【做法】

1. 臘腸隔水蒸 10 分鐘，切小粒，備用。

2. 糯米洗淨，瀝乾水分，加水用電飯煲煮熟。

3. 鑊內下油 2 湯匙，爆香乾蔥頭、蝦米及香菇，拌入臘腸粒炒香，關火，加入糯米飯及調味料拌勻，盛於蒸籠內。

4. 膏蟹洗淨，切塊，放於飯面，鋪上薑片，蒸 25 分鐘至膏蟹全熟，最後棄去薑片，灑上蔥粒即可上桌供食。

【 Method 】

1. Steam the sausages for 10 minutes. Dice finely. Set aside.

2. Rinse the glutinous rice. Drain. Pour in water. Cook in an electric rice cooker until done.

3. Heat 2 tbsps of oil in a wok. Stir fry shallot, dried shrimps and dried black mushrooms until fragrant. Add sausages and stir well. Turn off the heat. Put in the cooked glutinous rice and seasoning. Stir well. Transfer the resulting mixture into a steamer.

4. Rinse the crabs. Cut into pieces. Arrange over the rice mixture and place sliced ginger on top. Steam for 25 minutes until the crabs are done. Remove the sliced ginger. Sprinkle diced spring onion on top at last. Serve.

【小秘訣 • TIPS】

- 材料與糯米飯拌勻時，建議用筷子比湯匙好，糯米飯才不會拌得太糊。

- 煮糯米飯，水的份量是米的 70% 至 80% 已足夠。

- When you mix the glutinous rice with other ingredients in step 3, use a pair of chopsticks instead of a spatula. The glutinous rice won't be too mushy this way.

- When you cook glutinous rice, the ratio between rice and water should be 10 parts of rice to 7 or 8 parts of water.

炸腐球 Deep Fried
Inside-out
Stuffed
Beancurd Puffs

將豆卜反出來炸，
附着的那層薄豆腐
就是脆口的來源。
這個菜脆口不油
膩，吃過的朋友都
非常喜愛。

【材料】
豆腐卜 30 個
粟粉 2 湯匙（埋口用）

【餡料】
玉豆 200 克（切粒）
紅蘿蔔粒 100 克
香菇 4 隻（切粒）
蝦米碎半杯
肉碎半杯
木耳絲半杯
已浸粉絲 15 克（切小段，後下）
乾葱頭 3 粒（切片）
油 4 湯匙

【調味料 1】
蠔油 2 湯匙
生抽 2 湯匙
葱頭酥 1 湯匙
胡椒粉 1 茶匙
糖 1 茶匙

【調味料 2】
清雞湯半杯
麻油 2 湯匙
酒 3 湯匙
乾粟粉 2 湯匙（最後下）

【 Ingredients 】
30 beancurd puffs
2 tbsps cornstarch (for sealing the seam)

【 Filling 】
200 g white string beans (diced)
100 g diced carrot
4 dried black mushrooms (diced)
1/2 cup chopped dried shrimps
1/2 cup ground pork
1/2 cup shredded wood ear fungus
15 g mung bean vermicelli (soaked in water until soft, cut into short lengths, added at last)
3 cloves shallot (sliced)
4 tbsps oil

【 Seasoning 1】
2 tbsps oyster sauce
2 tbsps light soy sauce
1 tbsp deep-fried shallot
1 tsp ground white pepper
1 tsp sugar

【 Seasoning 2】
1/2 cup chicken stock
2 tbsps sesame oil
3 tbsps rice wine
2 tbsps cornstarch (added at last)

【做法】

1. 燒熱油 4 湯匙，下乾葱片爆香，加入香菇粒、蝦米及肉碎炒香，再加入玉豆及紅蘿蔔拌炒，最後拌入木耳絲及粉絲，灑下調味料 (1) 炒至香味撲鼻，加入調味料 (2) 炒透，最後灑入粟粉拌勻，取出，擱涼備用。

2. 豆卜用水略沖，抹乾水分，剝開一小口，翻轉豆腐卜，豆腐白膜在外。

3. 將餡料釀入豆卜內，於開口處抹上粟粉，豆卜皮相疊合上口。

4. 煎鑊內燒熱油 4 湯匙，豆卜開口處向下放入油鑊內，用中小火半煎炸至豆卜微黃及香酥，上桌時伴喼汁及五香椒鹽。

【Method】

1. Heat 4 tbsps of oil in a wok. Stir fry shallot until fragrant. Add dried black mushrooms, dried shrimps and ground pork. Stir until fragrant. Add string beans and carrot. Stir further. Lastly add wood ear fungus and mung bean vermicelli. Sprinkle with seasoning (1) and cook until fragrant. Add seasoning (2) and cook through. Sprinkle with cornstarch to thicken the filling. Set aside to let cool.

2. Rinse the beancurd puffs. Wipe dry. Make a small cut on them and turn them inside out.

3. Stuff the filling into the beancurd puffs. Smear some cornstarch on the seam. Overlap the skin of the beancurd puffs a little to seal the seam.

4. Heat 4 tbsps of oil in a wok. Place the stuffed beancurd puffs in the oil with the seam side down. Half-deep fry the beancurd puffs over medium-low heat until lightly browned and crispy. Serve with Worcestershire sauce and five-spice peppered salt on the side.

酸辣蝦仔

Spicy and Sour
Baby Shrimps

鮮甜的蝦，配上酸辣味道，美味自不待言，開胃得很，不但好下飯，伴意粉同吃更是絕配，冷吃也非常合適呢！

【材料】

蝦仔 450 克
番茄 1 個
紅辣椒 1 隻（切碎）
指天椒 1 隻（切碎）
蒜頭 3 粒（剁茸）
乾葱頭 3 粒（剁茸）
薑茸 1 湯匙

【調味料】

辣椒油、麻油各 1 茶匙
鹽、胡椒粉及雞粉各半茶匙
糖、生抽及紹酒各 1 湯匙
茄汁半湯匙
鎮江香醋 2 湯匙（後下）

【Ingredients】

450 g baby shrimps
1 tomato
1 red chilli (finely chopped)
1 bird eye chilli (finely chopped)
3 cloves garlic (grated)
3 cloves shallot (grated)
1 tbsp grated ginger

【Seasoning】

1 tsp chilli oil
1 tsp sesame oil
1/2 tsp salt
1/2 tsp ground white pepper
1/2 tsp chicken bouillon powder
1 tbsp sugar
1 tbsp light soy sauce
1 tbsp Shaoxing wine
1/2 tbsp ketchup
2 tbsps Zhenjiang black vinegar
(added at last)

【做法】

1. 蝦仔去殼、挑去腸，用生粉洗淨，抹乾。

2. 於番茄底部剠十字，放入滾水內煮約 10 秒，見外皮裂開即取出，去皮及去籽，切小丁。

3. 熱鑊下油，先爆香薑茸，再下乾葱茸、蒜茸、辣椒碎及指天椒碎爆至香味散發，放入蝦仁拌炒至轉成粉紅色，加入番茄及調味料再炒一會，於鑊邊澆入紹酒及醋即成，可配飯或意粉同吃。

【Method】

1. Shell the shrimps and devein. Rub caltrop starch generously on them. Rinse well and wipe dry.

2. Make a light crisscross incision on the bottom of the tomato. Blanch in boiling water for 10 seconds. Drain when the peel starts to come off. Peel it and seed it. Dice finely.

3. Heat oil in a wok. Stir fry ginger until fragrant. Then add shallot, garlic, red chilli and bird eye chilli. Stir until fragrant. Put in the shrimps and toss until they turn pink. Add tomato and seasoning. Stir briefly. Sizzle with Shaoxing wine and Zhenjiang black vinegar along the rim of the wok. Serve with rice or pasta.

【小秘訣 • TIPS】

若不用蝦仁，可改用鮮魷魚代替，色香味俱佳。

You may use squid instead of shelled shrimps as an equally tasty variation.

煎海鮮豆腐 Pan-fried
Seafood
Beancurd

想做一道老少皆宜、上桌得體、既美味又營養豐富的菜式，與家裏的小朋友一起動手，享受親子之樂吧！

這個菜最重要的秘訣是必須用手盡量擠乾豆腐，若含水分太多則難以成形，切時容易散開。另外，也建議放進冰箱冷藏，拍上麵粉煎後有微黃香脆的效果。

【 材 料 】
實豆腐 2 塊
馬鈴薯 2 個
鮮蝦半斤
魚柳肉 1 塊（約 150 克）
雞蛋 1 隻
葱粒半杯
麵粉 1 杯

【 伴 吃 】
生菜 3 棵
芝麻適量

【 魚蝦調味料 】
紹酒、胡椒粉、薑汁、麻油及
鹽各 1 茶匙

【 總調味料 】
粟粉 3 湯匙
番薯粉 3 湯匙
糖半茶匙
鹽及雞粉各 1 茶匙

【 Ingredients 】
2 cubes firm beancurd
2 potatoes
300 g fresh shrimps
1 piece of fish fillet (about 150 g)
1 egg
1/2 cup diced spring onion
1 cup flour

【 Side dish 】
3 lettuces
sesame seeds

【 Seasoning for fish and shrimps 】
1 tsp Shaoxing wine
1 tsp ground white pepper
1 tsp ginger juice
1 tsp sesame oil
1 tsp salt

【 Seasoning for minced beancurd mixture 】
3 tbsps cornstarch
3 tbsps sweet potato starch
1/2 tsp sugar
1 tsp salt
1 tsp chicken bouillon powder

【做法】

1. 豆腐置於隔篩內，灑上幼鹽 1 茶匙搓爛，盡量瀝乾水分。

2. 馬鈴薯蒸熟，搓成茸；雞蛋拂打成蛋液。

3. 蝦去殼、挑腸，與魚柳同切成小粒，加調味料醃 10 分鐘。

4. 所有材料及總調味料放入大碗內徹底拌勻。

5. 預備 8 吋 x 10 吋的糕盆，鋪上錫紙及抹上油，放入步驟（4）的豆腐混合物，蒸 20 分鐘，關火，取出。

6. 待冷後，放入冰箱待 3 小時或以上，倒扣碟上，切件，拍上少許麵粉，用少許油煎香。灑上芝麻及以生菜包裹，伴喼汁同吃。

【 Method 】

1. Place the beancurd in a strainer. Sprinkle with 1 tsp of table salt. Mash it through the strainer. Remove as much water as possible.

2. Steam the potatoes until done. Mash them. Whisk the egg.

3. Shell the shrimps and devein. Dice the shrimps and fish fillet. Add seasoning and mix well. Leave them for 10 minutes.

4. In a large mixing bowl, put in all ingredients. Add seasoning for minced beancurd mixture. Mix well.

5. Line a 8" x 10" baking tray with aluminium foil. Grease the foil. Pour in the minced beancurd mixture. Steam for 20 minutes. Turn off the heat.

6. Leave it to cool. Refrigerate for at least 3 hours. Turn the set beancurd out on a plate. Slice it and coat thinly in flour. Fry in a little oil until golden and crispy. Sprinkle with sesame seeds. Serve with lettuce for wrapping the fried beancurd and serve Worcestershire sauce as a dip.

【 小秘訣 • TIPS 】

● 宜選購街市售賣之散裝豆腐，因超市之盒裝豆腐較軟身，煎時較困難。

● 豆腐灑上幼鹽 1 茶匙，令豆腐滲出水分，與魚蝦拌勻後不致水分過多，口感佳！

● It's preferable to get bulk beancurd from wet market. Those pre-packed ones from supermarket are too soft and it is difficult for frying.

● Sprinkling 1 tsp of table salt on the beancurd helps draw out the moisture, so that the minced beancurd mixture will not be too watery at last.

蚵仔煎
Stir Fried Oyster with Duck Eggs

似乎想到福建菜，就想到「蚵仔煎」，可見其代表性。

蚵仔是家鄉盛產的海味，曬乾後更可做成不同菜餚的配搭，不説蚵仔煎的美味，單是營養價值已非常高呢！

【材 料】

蚵仔 400 克
番薯粉 6 湯匙
鴨蛋或雞蛋 4 隻
乾葱頭 3 粒（切片）
豬油或油 4 湯匙
芫茜 3 棵（切碎）

【調 味 料】

雞粉、糖、紹酒及麻油各 1 茶匙
胡椒粉半茶匙
魚露及沙茶醬各 1 湯匙

【 Ingredients 】

400 g baby oysters
6 tbsps sweet potato starch
4 duck or chicken eggs
3 cloves shallot (sliced)
4 tbsps lard or cooking oil
3 sprigs coriander (finely chopped)

【 Seasoning 】

1 tsp chicken bouillon powder
1 tsp sugar
1 tsp Shaoxing wine
1 tsp sesame oil
1/2 tsp ground white pepper
1 tbsp fish gravy
1 tbsp Sa Cha sauce

【蚵仔的清洗方法】

蚵仔買回後先放進冰箱，煮前才取出清洗。
為保新鮮，建議即日買即日吃。

1. 清洗蚵仔時，加粟粉或生粉 1 湯匙用手輕輕拌勻，放在水喉下沖洗（別開大水），重覆兩次。

2. 清洗時，徹底去除殼屑，為求吃得安全，最宜用手摸一遍，最後用清水洗淨，盛於漏勺上，瀝乾水分。

【做法】

1. 大碗內放入蚵仔及調味料拌勻（下鑊煎時，蚵仔瀝乾水分，與番薯粉拌勻）。

2. 鑊置於高火上燒至大熱，下豬油爆香乾葱片，放入蚵仔撥散煎香，見蚵仔表面粉漿呈微黃色，直接打入鴨蛋與蚵仔拌炒至香酥，最後灑入芫茜碎拌勻。上桌時，配泰國甜辣醬或魚露享用。

【小秘訣 • TIPS】

- 為了賣相美觀，通常將蚵仔煎煎成餅狀；但蚵仔與鴨蛋拌散煎香，口感更香酥可口。

- 清洗蚵仔時勿加鹽，以免蚵仔爆裂糊成一團。

- For better presentation, most restaurants serve this dish as an omelette or patty. But I prefer it to resemble scattered baby oysters coated in duck eggs. It tastes more flavourful and better this way.

- Do not add salt to the oysters when you rinse it. Otherwise, the oysters will turn mushy and you end up with a slimy mess.

【 Rinsing the baby oysters 】

Keep the baby oysters in the fridge. Rinse them right before you cook them. To ensure food safety and freshness, always cook and serve baby oysters on the same day you buy them.

1. To rinse them, add 1 tbsp of cornstarch or caltrop starch. Rub them with your hands well. Then rinse them under a slowly running tap. Repeat this step twice.

2. As you rinse them, also remove any broken shells on the oysters. For the best result, rub them with your hands again to double check. Rinse in tap water once more at last. Leave the oysters in a strainer until dry.

【 Method 】

1. Put the baby oysters and seasoning in a large mixing bowl. Stir well. Then drain the baby oysters again. Mix with the sweet potato starch and stir well.

2. Heat a pan over high heat until smoking hot. Add lard and stir fry shallot until fragrant. Put in the oysters and scatter them. Fry until the sweet potato starch on the oysters turns lightly browned. Beat in the duck eggs and stir quickly to coat oysters in the eggs. Fry until the eggs turn golden. Sprinkle with coriander and stir well. Serve with Thai sweet chilli sauce or fish gravy on the side.

蚵仔卷 Baby Oyster
Spring Rolls

【材料】
春卷皮 1 包（超市包裝）
番薯粉 3 湯匙
乾葱頭 1 湯匙（切片）

【餡料】
蚵仔 300 克（清洗方法見第 56 頁）
韭黃 1 杯（切段）
洋葱 1 杯（切幼絲）
銀芽 2 杯（洗淨、抹乾）
豬肉碎半杯

【調味料】
鹽及糖各 1 茶匙
沙茶醬 2 湯匙
粟粉 1 湯匙

【 Ingredients 】
1 pack frozen spring roll wrappers
(from supermarket)
3 tbsps sweet potato starch
1 tbsp sliced shallot

【 Filling 】
300 g baby oysters (refer to p.56 for rinsing method)
1 cup yellow chives (cut into short lengths)
1 cup shredded onion
2 cups mung bean sprouts (rinsed and wiped dry)
1/2 cup ground pork

【 Seasoning 】
1 tsp salt
1 tsp sugar
2 tbsps Sa Cha sauce
1 tbsp cornstarch

【做法】
1. 蚵仔清洗及瀝乾水分，加入紹酒 1 湯匙、魚露及胡椒粉各 1 茶匙、番薯粉拌勻。

2. 豬肉碎用生抽半湯匙醃好。

3. 燒滾大半鑊水，蚵仔放於漏勺，放入滾水內（滾水蓋過蚵仔），用筷子輕拌約 20 秒，原勺取出，瀝乾水分。

4. 熱鑊下油炒乾葱片，下豬肉碎炒香，加入韭黃、洋葱及銀芽略拌炒，灑入調味料拌勻，待涼備用。

5. 春卷皮鋪平，放上少量餡料，排上蚵仔捲好，用粟粉水封口，約做 15 卷。

6. 蚵仔卷放入溫油內，以中火炸至金黃香脆，吃時蘸喼汁或泰式甜辣醬。

【 Method 】
1. Rinse the baby oysters and drain. Stir in 1 tbsp of Shaoxing wine, 1 tsp of fish gravy, 1 tsp of ground white pepper and sweet potato starch.

2. Add 1/2 tbsp of light soy sauce to the ground pork. Mix well.

3. Boil over half a wok of water. Place the baby oysters into a strainer ladle. Transfer into the boiling water. Stir gently with chopsticks and cook for about 20 seconds. Drain.

4. Heat oil and stir fry shallot. Put in the ground pork. Add yellow chives, onion and bean sprouts. Stir briefly. Mix the seasoning. Set aside.

5. Lay flat a sheet of spring roll wrapper. Arrange filling and baby oysters on top. Roll it. Secure the seam with cornstarch solution. It makes 15 spring rolls.

6. Deep fry the spring rolls in warm oil over medium heat until crispy. Serve with Worcestershire sauce or Thai sweet and sour sauce as a dip.

香糟腩片

Braised Pork Belly
with Distillers' Grains

冰糖必須最後才放入，因冰糖的份量較多，而且豬肉切得薄，若早已下糖，腩肉容易燒焦。冰糖後下煮溶後，調至大火收汁，這時豬肉上色，帶一陣焦糖的香氣，更能突出此菜的色香味。

【材料】

花生 1 杯
腩肉 1 斤
薑 3 厚片
乾葱頭 6 粒
蒜頭 6 粒
芫茜 2 棵
油適量

【調味料】

紅糖 3 湯匙
紹酒半杯
生抽 4 湯匙
滾水 2 杯
碎冰糖 3 湯匙（後下）

【 Ingredients 】

1 cup peanuts
600 g pork belly
3 thick slices ginger
6 cloves shallot
6 cloves garlic
2 sprigs coriander
oil

【 Seasoning 】

3 tbsps red distillers' grains
1/2 cup Shaoxing wine
4 tbsps light soy sauce
2 cups boiling water
3 tbsps crushed rock sugar (added at last)

【做法】

1. 腩肉洗淨，切成 1 吋 x 1 1/2 吋長及半吋厚塊狀。

2. 薑、乾葱頭及蒜頭拍扁；芫茜切碎。

3. 鑊內下油小半鑊，冷油放入花生，用小火炸香約 30 分鐘，取出備用。

4. 預備另一個鑊，下油 1 湯匙爆香薑、乾葱頭及蒜頭，下腩肉煎至微黃，傾出多餘油分，下所有調味料（冰糖除外），加蓋小火燜煮 20 分鐘（期間拌炒一兩次以防燒焦），開蓋，加冰糖碎拌煮至冰糖溶化，轉中火煮至汁液收乾，最後加入炸花生拌勻，上碟時灑上芫茜碎即可。

【 Method 】

1. Rinse the pork belly. Cut into half-inch-thick pieces about 1" by 1.5" in sizes.

2. Gently crush the ginger, shallot and garlic. Set aside. Finely chop the coriander.

3. Add less than 1/2 a wok of oil. Put in the peanuts when the oil is still cold. Then turn on low heat and deep fry the peanuts for 30 minutes. Drain.

4. In another wok, heat 1 tbsp of oil and stir fry ginger, shallot and garlic until fragrant. Sear the pork belly pieces until lightly browned. Drain. Add seasoning (except the rock sugar). Cover the lid and simmer for 20 minutes over low heat. Stir the mixture once or twice throughout the cooking time to prevent it from burning. Stir in the rock sugar until it dissolves. Turn to medium heat to cook the sauce down. Lastly toss in deep fried peanuts. Mix well. Transfer to a serving plate and arrange chopped coriander on top. Serve

蹄筋煮蘿蔔

Braised Sinews
with White Radish

瑤柱及魷魚有提鮮作用，蹄筋吸收瑤柱及魷魚的味道，更鮮味好吃。你可用鮮蝦、蟹肉或蜆肉代替魷魚，味道俱佳。

浸發的蹄筋質感爽口、不油膩，與燜煮的蹄筋口味各異，是較特別的吃法。

【材料】
浸發豬蹄筋 8 條
（浸發方法參考第 91 頁）
瑤柱 2 粒
白蘿蔔 600 克
小魷魚 300 克
清雞湯 2 杯
水 1 杯
蒜頭及乾葱頭各 1 粒（切片）
薑 1 片

【調味料】
酒及麻油各 2 茶匙

【做法】
1. 蘿蔔去皮，切粗條；瑤柱用水浸 30 分鐘，去硬枕，撕成粗絲；每條蹄筋切為兩件。

2. 小魷魚洗淨，原隻保留備用。

3. 燒熱油 1 湯匙，爆香薑、乾葱及蒜片，傾入清雞湯及水煮滾，下瑤柱及蘿蔔用中火加蓋煮至蘿蔔熟透，加入蹄筋再煮 2 至 3 分鐘，下調味料拌勻。

4. 蘿蔔、瑤柱及蹄筋放在深碟內，小魷魚放入鑊內之湯汁煮滾及熟透，連湯汁傾於蘿蔔上即可。

【小秘訣 • TIPS】
蹄筋含豐富骨膠原，具美化肌膚的作用。

Pork sinews are rich in collagen which enhances skin texture.

【 Ingredients 】
8 pieces re-hydrated pork heel sinews
(please refer to p.91 for re-hydration method)
2 dried scallops
600 g white radish
300 g baby squids
2 cups chicken stock
1 cup water
1 clove garlic (sliced)
1 clove shallot (sliced)
1 slice ginger

【 Seasoning 】
2 tsps rice wine
2 tsps sesame oil

【 Method 】
1. Peel the white radish and cut into thick strips. Soak the dried scallops in water for 30 minutes. Remove the tough tendon. Tear them into fine shreds. Cut each piece of sinew into halves.

2. Rinse the baby squids and keep them in whole.

3. Heat 1 tbsp of oil in a wok. Fry the ginger, shallot and garlic until fragrant. Add chicken stock and water. Bring to the boil. Add dried scallops and radish. Turn to medium heat and cover the lid. Cook until radish is done. Put in the sinews and cook for 2 to 3 more minutes. Add seasoning. Mix well.

4. Arrange the radish, dried scallops and sinews in a deep dish. Cook the baby squids in the sauce until done. Pour the mixture over the radish and serve.

鹽酥拼盤

Deep Fried
Snack Platter

這個拼盤最宜看電視時作零食，看世界杯足球賽時，這個小食能不受歡迎嗎？

杏鮑菇營養高，帶甜又爽口，用微波爐先叮熟，以便快炸，保持菇的鮮汁不會流失。

【材料】
雲吞皮 70 克
皮蛋 2 隻
杏鮑菇約 200 克
葱碎、蒜茸及辣椒碎各 1 湯匙

【調味料】
生抽 1 湯匙
糖及薑米各 1 茶匙

【炸粉】
五香椒鹽半湯匙（做法參考第 79 頁）
自發粉 1/4 杯
番薯粉 3/4 杯

【Ingredients】
70 g wonton wrappers
2 thousand-year eggs
200 g king oyster mushrooms
1 tbsp chopped spring onion
1 tbsp grated garlic
1 tbsp chopped red chillies

【Seasoning】
1 tbsp light soy sauce
1 tsp sugar
1 tsp finely diced ginger

【Deep-frying flour mix】
1/2 tbsp five-spice peppered salt (please refer to p.79 for method)
1/4 cup self-raising flour
3/4 cup sweet potato starch

【小秘訣 • TIPS】

- 材料下油鍋前才沾上炸粉,靜置於室溫 10 分鐘,讓其「反潮」,炸粉才不容易散落熱油內,變得混濁。

- 炸好的食物從油鍋取出後才關火,以免吸油太多。炸物建議先放於廚房紙,吸乾油分才排於碟上。

- 炸皮蛋要用猛火快炸,因皮蛋不宜炸太久,容易軟化。

- Before you deep fry the ingredients, coat them in the deep-frying flour mix and leave them for 10 minutes first. The flour mix will adhere to the ingredients more securely this way and won't darken the oil as much.

- Turn off the heat only after you've taken the food out of the oil. Otherwise, the deep-fried goods will be greasy. Leave them on paper towel to absorb excess oil first. Then transfer onto a serving plate.

- Make sure you deep fry the thousand-year egg over high heat very quickly. It tends to melt when fried for too long.

【做法】

1. 炸粉拌勻,放在碟上。

2. 雲吞皮切成粗條,用溫油慢火炸至金黃,取出,灑上少許五香椒鹽放在碟上。

3. 每隻皮蛋切成 4 塊,沾上炸粉,用猛火炸至外皮微黃,取出,排於碟上。

4. 於杏鮑菇底部剠十字,放入微波爐用高火叮 2 分鐘,取出,撕成粗條,加入調味料,沾上炸粉,用中火炸至金黃,盛起。

5. 鑊內不用下油,用小火爆香蒜茸、葱碎及辣椒碎,關火,下杏鮑菇及五香椒鹽 1 茶匙拌勻,排上碟享用。

【Method】

1. Mix the deep-frying flour mix first. Arrange on a flat plate.

2. Cut the wonton wrappers into thick strips. Deep fry in warm oil over low heat until golden. Drain. Sprinkle some five-spice peppered salt over. Arrange on a serving plate.

3. Cut each thousand-year egg along the length into quarters. Coat them in the deep-frying flour mix. Deep fry over high heat until lightly browned. Arrange on a serving plate.

4. Make a crisscross cut on the base of the king oyster mushrooms. Heat in a microwave oven over high power for 2 minutes. Tear into thick strips. Add seasoning. Then coat them in the deep-frying flour mix. Deep fry over medium heat until golden. Drain.

5. With the remaining oil in the wok, fry the grated garlic, spring onion and chilli until fragrant. Turn off the heat. Toss in the fried king oyster mushrooms and add 1 tsp of five-spice peppered salt. Mix well. Arrange on the serving plate. Serve.

Carp Belly
in Red Distillers'
Grains Sauce

紅糟魚

這是福建名菜，逢年過節，家裏都愛做，顏色紅紅的很喜氣。紅糟魚配粥、飯或下酒俱皆，也常用作冷盤。

紅糟乃是酒糟，故有濃濃酒香，對身體有益，可降膽固醇及令血管軟化，是近年流行的食材。紅糟汁可醃排骨、豬肉、雞或蝦等，炸食或蒸皆宜，請發揮你的創意烹調吧！

【小秘訣 • TIPS】

- 紅糟可於新三陽上海南貨店購買。

- 魚不要切得太薄，約1吋厚度為佳，以免炸時容易散碎及太乾，而且醃製後食用，能保持完整塊狀。

- 紅糟魚放在冰箱可保鮮 7 至 10 日，魚件醃得更入味、更好吃。

- You can get red distillers' grains from Shanghainese grocery stores, such as Shanghai New Sam Yung Market.

- Do not slice the fish too thinly. Each slice should be about 1 inch thick. Thinly sliced fish tends to break down into bits and pieces in the frying process. It also tends to be dry and tough after being fried.

- This dish lasts well in the fridge for 7 to 10 days. The longer the fish is marinated in the sauce, the tastier it is.

【材料】
紅糟 2 樽（共 450 克）
鯇魚腩 2 斤（切厚片）
油 1 瓶（900 毫升）
蒜茸 3 湯匙

【醃料】
薑 4 片（拍扁）
葱 4 條（切段、拍扁）
蒜頭 4 粒（切片）
胡椒粉半湯匙
鹽、糖及麻油各 1 湯匙

【調味料】
紹酒 1/4 杯
生抽半杯
糖 3 湯匙
油 3 湯匙
鹽 1 茶匙

【 Ingredients 】
2 bottles red distillers' grains (450 g in total)
1.2 kg grass carp belly (sliced thickly)
900 ml oil
3 tbsps grated garlic

【 Marinade 】
4 slices ginger (gently crushed)
4 sprigs spring onion (cut into short lengths; gently crushed)
4 cloves garlic (sliced)
1/2 tbsp ground white pepper
1 tbsp salt
1 tbsp sugar
1 tbsp sesame oil

【 Sauce 】
1/4 cup Shaoxing wine
1/2 cup light soy sauce
3 tbsps sugar
3 tbsps oil
1 tsp salt

【做法】

1. 鯇魚件與醃料拌勻醃 2 小時（醃一晚更佳）。

2. 預備紅糟汁：鑊內下油 3 湯匙，油熱時爆香蒜茸，傾入紅糟用小火略炒，加入調味料煮 10 分鐘，期間不斷翻動（由於紅糟搶火，容易燒焦），關火，盛起備用。

3. 鑊內燒熱油 1 瓶，魚件分兩批放入熱油，用中小火炸至金黃色，盛起，瀝乾油分。

4. 在深盤內，放入紅糟汁，排上已炸魚塊，再鋪上紅糟汁及魚塊，如此類推，待涼，加蓋，放入冰箱冷藏，隔日取出，伴檸檬作為冷盤，或蒸熱享用。

【 Method 】

1. Add marinade to the sliced grass carp. Mix well and leave it for at least 2 hours. (It tastes even better if you leave it overnight).

2. To make the red distillers' grains sauce, heat 3 tbsps of oil in a wok. Stir fry garlic until fragrant. Pour in the red distillers' grain. Stir briefly over low heat. Add sauce ingredients and cook for 10 minutes. Keep stirring continuously. Turn off the heat. Set aside.

3. In another wok, heat up 900 ml of oil. Deep fry the sliced carp in two batches over medium-low heat until golden. Set aside. Drain off the oil.

4. In a deep tray, pour in a layer of the red distillers' grains sauce. Arrange the deep fried carp pieces in the sauce. Pour on another layer of the sauce. Top with another layer of carp pieces. Repeat this step until all carp pieces are used up. Leave them to cool. Cover the tray. Refrigerate overnight. Serve as a cold appetizer with lemon on the side. Or steam it to reheat to serve hot.

滷水拼盤

Marinated
Meat
Platter

希望大家別見材料多而放棄不做，因弄滷水是不會失敗，而且滷汁可滷製很多食材，作為自己享用或送給親友品嘗，若吃不完可用保鮮紙包好，冷藏分三數日食用，煮湯麵或炒菜，更可做成沙律，吃法多樣化，故滷水是過年時必備的食品。

滷水的食材數之不盡，可按自己的愛好取材：

吃素者可選雲耳、豆乾、筍、高麗菜、雞蛋、麵筋等。

下酒的可選鵝掌、鵝翼、鵝腸、鴨舌、鴨腎、乳鴿、墨魚、整條腩肉、牛肚、牛筋等。

腩肉切小丁，滷好後放在麵或飯上，配上滷蛋及雲耳，就是很多人喜愛的魯肉飯及肉燥麵了。試試吧！福建滷水是很好吃的！

【滷水材料】
八角 6 粒
月桂葉 6 片
桂皮 1 小塊
甘草 4 片
陳皮 1/4 角
白胡椒粒 1 湯匙
花椒粒 2 湯匙

【爆香材料】
乾葱頭及蒜頭各 6 粒
（拍扁）
薑 6 厚片
葱 3 條（切段）
紅辣椒 1 隻（切段）

【調味料】
紹酒及生抽各半杯
蠔油及冰糖各 1/4 杯
老抽及雞粉各 1 湯匙
五香粉及魚露各半湯匙
沙薑粉 2 湯匙
水 4 杯

【滷水蘸汁 1】
蒜茸 1 湯匙
老抽 2 湯匙
糖、麻油、辣椒油、雞湯、
醋、芫茜末或葱末各 1 湯
匙

【滷水蘸汁 2】
糖及白醋各 2 湯匙
青檸檬汁及魚露各 2 湯匙
紅辣椒半隻（切碎）
蒜茸 1 湯匙

【 Marinade spices 】
6 cloves star anise
6 bay leaves
1 small piece cassia bark
4 slices liquorice
1/4 dried tangerine peel
1 tbsp white peppercorns
2 tbsps Sichuan peppercorns

【 Aromatics 】
6 cloves each of shallot
and garlic (crushed)
6 thick slices ginger
3 sprigs spring onion (sectioned)
1 red chilli (cut into short sections)

【 Marinade ingredients 】
1/2 cup Shaoxing wine
1/2 cup light soy sauce
1/4 cup oyster sauce
1/4 cup rock sugar
1 tbsp dark soy sauce
1 tbsp chicken bouillon powder
1/2 tbsp five-spice powder
1/2 tbsp fish gravy
2 tbsps spice ginger powder
4 cups water

【 Dipping sauce 1 】
1 tbsp grated garlic
2 tbsps dark soy sauce
1 tbsp sugar
1 tbsp sesame oil
1 tbsp chilli oil
1 tbsp chicken stock
1 tbsp vinegar
1 tbsp finely chopped coriander or
grated spring onion

【 Dipping sauce 2 】
2 tbsps sugar
2 tbsps vinegar
2 tbsps lime juice
2 tbsps fish gravy
1/2 red chilli (finely chopped)
1 tbsp grated garlic

【 食材 的 處理 方法 】

豬耳： 飛水，用小鉗拔淨雜毛。

豬肚： 通常買回來時是兩層的，用剪刀
剪開成一大塊，用鹽及生粉洗刷
多次。

豬大腸： 反轉內層向外，用鹽及生粉洗刷
兩次，飛水後再翻轉。

鴨舌： 飛水，洗乾淨。

鵝翼： 飛水，用小鉗拔清雜毛。

【 Preparing the frozen meat 】

Pork ear: Blanch in boiling water. Pluck off the hair with a pair of tweezers.

Pork tripe: It usually comes in two layers. Cut open it with a pair of scissors. Then rub salt and caltrop starch on both the insides and outsides. Rinse well. Rub salt and caltrop starch on it again and rinse. Do it repeatedly.

Pork chitterlings: Turn them inside out. Rub salt and caltrop starch on them. Rinse well. Rub salt and caltrop starch on them for the second time. Rinse. Blanch and turn them inside out again.

Duck tongues: Blanch them in boiling water and rinse well.

Goose wings: Blanch them in boiling water and pluck the fine hair with a pair of tweezers.

* 以下的材料選用急凍食品製作 。

材料 ：	滷水時間 ：
鵝翼 5 隻	90 分鐘
豬大腸 2 條	75 至 90 分鐘
豬肚 1 個	90 分鐘
豬耳 1 個	60 分鐘
鴨舌 1 磅	20 分鐘

已浸發香菇及刺參，用滷水汁 1 杯及水 1 杯滷製 15 分鐘。

*Marinating time for the frozen meat

5 goose wings:	90 minutes
2 pork chitterlings:	75 to 90 minutes
1 pork tripe:	90 minutes
1 pork ear:	60 minutes
450 g duck tongues:	20 minutes

To marinate dried black mushrooms or sea cucumbers that have been re-hydrated, add 1 part of water to 1 part of marinade. Cook them for 15 minutes.

【做法】

1. 大煲內燒滾水，放入薑3片、葱3條（切段）及酒1湯匙，逐一加入食材飛水5至10分鐘，取出，洗淨及抹乾備用（飛水即是在滾水內烚一會，去除血腥味及雜質）。

2. 瓦鍋內燒熱油3湯匙，下爆香材料用小火炒至香氣撲鼻，加入所有滷水材料同炒，傾入水及所有調味料煮滾，放入已處理的食材用小火滷製，保持滷汁煮滾（不同的食材，滷製時間有別）。

3. 切件上桌，配滷水蘸汁品嘗。

【 Method 】

1. Boil the water in a large pot. Then add 3 slices of ginger, 3 sprigs of spring onion and 1 tbsp of rice wine. Blanch the meat ingredients separately for 5 to 10 minutes. Drain. Rinse well and wipe dry. The purpose of blanching is to remove the blood and impurities in the meat.

2. Heat 3 tbsps of oil in a clay pot. Stir fry the aromatics over low heat until fragrant. Put in all marinade spices. Stir briefly. Pour in water and the marinade ingredients. Bring to the boil. Put in the blanched meat ingredients and cook them according to the specified time. *(Keep the marinade in a gentle simmer throughout the cooking process).

3. Slice the meat and serve the dipping sauces on the side.

【小秘訣 • TIPS】

若瓦鍋盛不下所有食材及配料，建議分批滷製。

If you can't fit all ingredients into a clay pot, do it in batches.

燴肚片花菇 刺參

Marinated Pork Tripe,
Mushrooms and
Sea Cucumber

【材料】

滷水豬肚 350 克

滷水香菇 8 隻

滷水海參 2 條（約 200 克）

黃芽白 350 克

蒜茸 1 湯匙

＊滷製方法參考第 72 及 73 頁。

【芡汁（拌勻）】

生粉半湯匙

滷水汁 1 杯

（若滷水汁太濃稠，可改用滷水汁半杯與清雞湯半杯拌勻）

【做法】

1. 豬肚及香菇切片；海參切大段。

2. 黃芽白洗淨，切粗絲，用鹽油水炒熟，取出，瀝乾菜汁。

3. 深碗內相間地排上花菇及豬肚，黃芽白填滿中間，倒扣在盤中，海參圍邊，頂端放香菇一朵，大火蒸 10 分鐘。

4. 燒熱油 1 湯匙，爆香蒜茸，勾芡，澆於豬肚花菇上即成。

【 Ingredients 】

350 g marinated pork tripe

8 marinated dried black mushrooms

2 marinated sea cucumbers (about 200 g)

350 g Napa cabbage

1 tbsp grated garlic

* Refer to p.72-73 for the marinating method.

【 Thickening glaze (mixed well) 】

1/2 tbsp caltrop starch

1 cup marinade for marinating meat

(if the marinade is too thick, use 1/2 cup of marinade and add 1/2 cup of chicken stock.)

【 Method 】

1. Slice the pork tripe and dried black mushrooms. Cut the sea cucumber into large chunks.

2. Rinse the cabbage and shred coarsely. Stir fry the cabbage in a little oil, water and salt until done. Drain well.

3. In a deep dish, arrange dried black mushrooms and pork tripe in alternate layers along the rim. Then fill the centre with the cooked cabbage. Turn them out neatly onto a serving plate. Arrange the sea cucumber along the rim of the plate. Put a whole dried black mushroom on top. Steam for 10 minutes.

4. Heat 1 tbsp of oil in a wok. Stir fry grated garlic until fragrant. Pour in the thickening glaze and bring to the boil. Drizzle over the steamed pork tripe and mushrooms. Serve.

茄子大腸煲 Marinated Pork Chitterlings and Eggplant in Clay Pot

【 材料 】

滷水大腸 300 克（做法見第 72 及 73 頁）
茄子 400 克
薑 2 厚片
紅辣椒 1 隻
蒜頭及乾葱頭各 3 粒
葱 2 條（切段）

【 調味料 1 】

醋、生抽、酒、麻油、糖各 1 湯匙

【 調味料 2 】

清雞湯半杯
金不換 2 杯
蒜茸 2 湯匙

【 做法 】

1. 茄子洗淨，切條；紅辣椒去籽、切塊；乾葱頭、蒜頭及薑拍扁。大腸切小段。

2. 熱鑊燒油 3 湯匙，下茄子煎香，盛起。

3. 燒熱瓦煲，下油 1 湯匙，下乾葱頭、蒜頭、薑及葱段爆至金黃色，放入滷水大腸、紅辣椒及茄子拌炒，下調味料 (1) 拌勻炒香，最後加入清雞湯及金不換拌勻，加蓋燜煮 5 分鐘，上桌前拌入蒜茸即成。

【 Ingredients 】

300 g marinated pork chitterlings (refer to p.72-73 for method)
400 g eggplant
2 thick slices ginger
1 red chilli
3 cloves garlic
3 cloves shallot
2 sprigs spring onion (cut into short lengths)

【 Seasoning 1 】

1 tbsp vinegar
1 tbsp light soy sauce
1 tbsp rice wine
1 tbsp sesame oil
1 tbsp sugar

【 Seasoning 2 】

1/2 cup chicken stock
2 cups Thai basil
2 tbsps grated garlic

【 Method 】

1. Rinse the eggplant. Cut into strips. Set aside. Seed the red chilli and cut into pieces. Gently crush the shallot, garlic and ginger. Cut the chitterlings into short lengths.

2. Heat 3 tbsps of oil in a wok. Put in the eggplant and fry until lightly browned. Set aside.

3. Heat a clay pot. Add 1 tbsp of oil. Stir fry the shallot, garlic, ginger and spring onion until golden. Put in the marinated chitterlings, chilli and eggplant. Stir well. Add seasoning (1) and stir fry until fragrant. Add chicken stock and Thai basil. Stir well. Cover the lid and simmer for 5 more minutes. Stir in the grated garlic at last and serve.

五香雞卷

Five-spice Pork Belly Rolls

這是很多福建同鄉的必煮菜式，雖名為雞卷，其實是豬肉卷，也有稱名「五香」是宴客必備菜式。

【 材 料 】

　豬腩肉 250 克
　馬蹄 6 粒
　洋葱半個
　蒜白及葱白各半杯
　乾葱茸及蒜茸各 1 湯匙
　豬網油 4 至 5 塊
　生粉適量

【 醃 料 】

　雞粉、胡椒粉、糖及鹽各 1 茶匙
　麻油及五香粉各 2 茶匙
　生抽 1 湯匙
　馬蹄粉或番薯粉 4 湯匙
　蛋黃 1 隻

【 五 香 椒 鹽 材 料 】

　胡椒粉 2 湯匙
　鹽 3 茶匙
　五香粉及雞粉各 2 茶匙
　肉桂粉及咖喱粉各半茶匙
　＊ 拌勻

【 Ingredients 】

　250 g pork belly
　6 water chestnuts
　1/2 onion
　1/2 cup white parts of Chinese leeks
　1/2 cup white parts of spring onion
　1 tbsp chopped shallot
　1 tbsp grated garlic
　4 to 5 pieces pork caul fat
　caltrop starch

【 Marinade 】

　1 tsp chicken bouillon powder
　1 tsp ground white pepper
　1 tsp sugar
　1 tsp salt
　2 tsps sesame oil
　2 tsps five-spice powder
　1 tbsp light soy sauce
　4 tbsps water chestnut starch or
　sweet potato starch
　1 egg yolk

【 Five-spice peppered salt 】

　2 tbsps ground white pepper
　3 tsps salt
　2 tsps five-spice powder
　2 tsps chicken bouillon powder
　1/2 tsp ground cinnamon
　1/2 tsp curry powder
　＊ mixed well

【 做法 】

1. 洋葱切碎;馬蹄去皮,洗淨,放入膠袋內拍成粗粒。

2. 豬腩肉細切粗剁,加入醃料醃 1 小時,加入其他材料(生粉除外)拌勻。

3. 豬網油請肉販清洗乾淨,回家後用鹽水及生粉水清洗兩次,以清水洗淨,抹乾水備用。

4. 豬網油鋪平,排入豬肉材料,捲起,用生粉水埋口,蒸 15 分鐘至熟(以上材料約可做 4 至 5 條)。

5. 肉卷拍上少許生粉,下油鑊用小火慢炸至表面微黃,轉大火續炸 10 秒至金黃香酥,切厚片上碟,蘸五香椒鹽、蒜茸醋或辣椒醬享用。

【 Method 】

1. Chop the onion and set aside. Peel the water chestnuts and rinse well. Place them into a plastic bag and crush them coarsely.

2. Cut the pork belly into small pieces and then chop coarsely. Add marinade and mix well. Leave it for 1 hour. Then add all other ingredients (except the caltrop starch). Mix well.

3. Ask the butcher to clean the caul fat for you. Then rinse them again with salted water and caltrop starch solution twice. Rinse well with water. Wipe dry.

4. Lay flat a piece of caul fat. Then arrange the pork filling over it. Roll it up like a spring roll. Seal the seam with some caltrop starch solution. Steam for 15 minutes until done. The amounts listed here make about 4 to 5 rolls.

5. Then lightly coat the caul fat rolls in dry caltrop starch. Fry in oil over low heat until lightly browned. Turn to high heat and fry for 10 seconds until golden. Slice and serve on a plate. Serve with five-spice peppered salt, garlic vinegar or chilli sauce on the side as a dip.

【 小秘訣 • TIPS】

- 五香椒鹽適用蘸吃任何酥炸食物,預備多一點存放於玻璃樽。

- 於過年時節,多弄些肉卷放於冰箱保鮮,方便隨時取出炸食。若不用豬網油,可改用腐皮。

- 肉餡也可做成肉丸炸吃,美味可口。

- You can make a bulk volume of five-spice peppered salt and store it in an airtight container. It goes well with all kinds of deep-fried food.

- Around Chinese New Year, you may make more of the pork rolls and store them in the fridge. Just deep fry them before serving. If you don't like caul fat, you may wrap the filling in beancurd skin instead.

- Just the filling itself also tastes great. You can deep fry the filling into meatballs for a tasty treat.

Soy Marinated
Pork Belly with
Dried Cuttlefish

墨魚乾魯肉

魯肉即「滷肉」，也稱為燜肉，燜五花腩本就美味，加上墨魚乾更別有風味，那種鮮香，沒有吃過是體會不到的。若感墨魚乾較硬，可用鮮墨魚同燜，吃了回味無窮！

81
Entrees
主菜

【材料】

腩肉 600 克
墨魚乾 120 克
薑 3 厚片
乾葱頭 8 粒
蒜頭 8 粒
八角 4 粒
花椒 1 湯匙
滾水 2 杯
浸墨魚乾水 1 杯

【調味料】

酒半杯
蠔油半杯
老抽 4 湯匙
冰糖、胡椒粉、五香粉及雞粉
各 1 湯匙

【 Ingredients 】

600 g pork belly
120 g dried cuttlefish
3 thick slices ginger
8 cloves shallot
8 cloves garlic
4 cloves star anise
1 tbsp Sichuan peppercorns
2 cups boiling water
1 cup soaking water for the dried cuttlefish

【 Seasoning 】

1/2 cup rice wine
1/2 cup oyster sauce
4 tbsps dark soy sauce
1 tbsp rock sugar
1 tbsp ground white pepper
1 tbsp five-spice powder
1 tbsp chicken bouillon powder

【做法】

1. 墨魚乾浸水 15 分鐘，去皮洗淨，用水蓋過面浸 2 小時（浸墨魚乾水留用），切成 3/4 吋 x 1 1/2 吋長塊。

2. 腩肉洗淨，刮淨毛，切成 2 吋 x 2 吋大塊。

3. 薑、乾葱頭及蒜頭拍扁；八角及花椒洗淨，八角用刀背拍成小塊。

4. 薑、乾葱頭、蒜頭、花椒及八角放進熱油，爆至香味四溢，加入腩肉及墨魚乾用大火煎至微黃，下調味料炒香後，注入水及浸墨魚乾水共 3 杯，水滾後轉放砂煲內，用小火慢煮燜 1 1/2 小時即可。

【 Method 】

1. Soak the dried cuttlefish in water for 15 minutes. Peel off the purple skin. Rinse well. Then soak it in water again for 2 hours (with enough water to cover it). Set aside the soaking water for later use. Cut into rectangular pieces about 3/4" by 1 1/2".

2. Rinse the pork belly. Scrape off the hair thoroughly. Cut into large chunks about 2" by 2".

3. Gently crush the ginger, shallot and garlic. Set aside. Rinse the star anise and Sichuan peppercorns. Gently crush the star anise into bits with the flat side of a knife.

4. Heat a wok and stir fry ginger, shallot, garlic, Sichuan peppercorns and star anise until fragrant. Sear the pork belly and dried cuttlefish over high heat until lightly browned. Add seasoning and stir until fragrant. Pour the soaking water for the dried cuttlefish first. Then pour in boiling water to make up to 3 cups of liquid. Bring to the boil over high heat. Transfer into a clay pot. Simmer over low heat for 1 1/2 hours. Serve.

【小秘訣 • TIPS】

- 燜食物時應將料頭（薑、乾蔥頭及蒜頭）逐一下鑊爆炒，通常先放入薑，炒至聞到薑香味，再下乾蔥頭炒一會，聞到乾蔥頭的香氣後，再下蒜頭，如此類推。

- 若烹調急凍食品，必須徹底解凍及洗淨，抹乾後才加調味料，加入酒可去除雪味。

- 調味料內沒加入生抽，因墨魚乾帶鹹味。若不用墨魚乾，可用海參燜煮，但需加入生抽 2 湯匙。

- 建議用小火燜煮，時間足夠，腩肉才好吃。若用大火燜，肉質容易變得乾硬。

- 燜煮時如需加水，宜傾入滾水，以保持食物的溫度。用冰糖調味較用砂糖好味。

- Before braising any food, stir fry the aromatics one after another first. Usually, I'd start with the ginger. When I can smell the fragrance of ginger, I'd add shallot. When I can smell the shallot, I'd add the garlic and so forth.

- Before you cook any frozen food, make sure you thaw it thoroughly and rinse well. Wipe dry before marinate or season it. Adding a splash of wine helps remove the unpleasant taste in most frozen food.

- I did not use any light soy sauce in the seasoning because the dried cuttlefish is salty in taste. You may also use sea cucumber instead of dried cuttlefish, in which case you should add 2 tbsps of light soy sauce to season it.

- Always braise pork belly over low heat for extended period. It takes time to pick up the flavour and for the fat to turn melty. If you cook it quickly over high heat, the meat will be dry and tough.

- When you braise food for a long period of time, check from time to time if it's drying up. Pour in boiling water in case not much liquid is left. That would keep the braising temperature more or less constant. I used rock sugar because it tastes better than white sugar. It also softens the meat somehow.

福建 • 湯羹
FUJIAN SOUPS

一碗媽媽的熱湯，
細味溫暖綿密的感覺，
享受回家的幸福！

A bowl of piping hot soup

The long missed flavour of mom's cooking

A taste of warmth and care

The joy of coming home

麵線 Flour Vermicelli in Thick Broth

麵線是家鄉人從小到大都愛吃的，是正餐也是小點，在東南亞、台灣、福建都很熱門流行。

配麵線的食材多不勝數，可用大腸、豬紅、蝦仁、魷魚、貢丸、魚丸等。若是生日，會吃豬肝及腰子麵線。

【材料】

蚵仔 300 克
蟶子（連殼）600 克
麵線 300 克
上湯 10 杯
芫茜 3 棵（切碎）
蒜茸 1 杯
葱粒適量
番薯粉 1 1/2 杯

【湯調味料】

沙茶醬及香菇蠔油各 2 湯匙
生抽及鎮江烏醋各 3 湯匙
糖、雞粉及胡椒粉各半湯匙

【蚵仔及蟶子調味料】

酒 1 湯匙
麻油 1 湯匙

【勾芡】

番薯粉半杯
水 1 杯
* 拌勻

【 Ingredients 】

300 g baby oysters
600 g razor clams (in shells)
300 g flour vermicelli
10 cups premium stock
3 sprigs coriander (finely chopped)
1 cup grated garlic
diced spring onion
1 1/2 cups sweet potato starch

【 Seasoning for the soup 】

2 tbsps Sa Cha sauce
2 tbsps mushroom oyster sauce
3 tbsps light soy sauce
3 tbsps Zhenjiang black vinegar
1/2 tbsp sugar
1/2 tbsp chicken bouillon powder
1/2 tbsp ground white pepper

【 Seasoning for baby oysters and razor clams 】

1 tbsp rice wine
1 tbsp sesame oil

【 Thickening glaze 】

1/2 cup sweet potato starch
1 cup water
* mixed well

一品窩　Premium
　　　　　Gourmet Soup

【材料】

鮮雞半隻
排骨 200 克
瑤柱 5 粒
急凍豬肚 300 克
乾豬蹄筋 6 條
浸發大海參 1 隻
香菇 4 隻
美國急凍螺頭 2 個
金華火腿 4 片
薑 3 厚片
蒜頭 6 粒
水約 6 杯（視乎燉盅大小而定）

【調味料】

酒 1 湯匙

【 Ingredients 】

1/2 chicken
200 g pork ribs
5 dried scallops
300 g frozen pork tripe
6 pieces dried pork heel sinews
1 dried sea cucumber (re-hydrated)
4 dried black mushrooms
2 U.S. frozen shelled conches
4 slices Jinhua ham
3 thick slices ginger
6 cloves garlic
about 6 cups water (depending on
the size of the double-steamer)

【 Seasoning for the soup 】

1 tbsp rice wine

【 小秘訣 • TIPS】

* 豬蹄筋用冷油才能炸透，另下油鍋時蹄筋必須用乾布抹乾，才不會濺起熱油。
* 建議多預備豬蹄筋，炸透及浸發後放入冰箱冷藏，無論燜煮及煲湯均可大派用場。
* You have to put in the sinews while the oil is cold for them to be deep fried thoroughly. Also wipe them completely dry before deep frying. Otherwise, the hot oil may splatter.
* You may prepare more pork heel sinews than you need. Just deep fry them thoroughly and re-hydrate them. Keep them in the freezer. They will be ready to use for braising or in soups.

【做法】

1. 雞洗淨，剁塊，與排骨一同飛水，洗淨。

2. 香菇浸軟，洗淨，切粗條。瑤柱洗淨，去掉硬枕。海參洗淨，切粗圈。

3. 豬肚用鹽及生粉洗擦乾淨。滾水內放少許葱、薑及酒，放入豬肚飛水 5 分鐘，取出，洗淨，切粗條。

4. 螺頭用牙刷擦淨，剪去腸，放入滾水內飛水 1 分鐘，取出，切厚片。金華火腿飛水，去油味。

5. 乾豬蹄筋用乾布抹淨，放入冷油內，用最小火炸約 30 分鐘，期間用筷子撥動，見豬蹄筋澎漲、有水泡及呈深金黃色時，盛起，浸於水內，擠出多餘油分，洗淨及擠水三數次，每條豬蹄筋切為兩段。

6. 燉盅內注入滾水，下薑片、蒜頭、雞、排骨、豬肚、香菇、瑤柱、火腿及螺頭燉 2 小時，加入豬蹄筋及海參再燉 30 分鐘，最後下調味料即可享用。

【 Method 】

1. Rinse the chicken and chop into pieces. Blanch chicken and ribs in boiling water together. Rinse well.

2. Soak the dried black mushrooms in water until soft. Rinse well and cut into thick strips. Set aside. Rinse the dried scallops. Remove the tough tendons and set them aside. Rinse the sea cucumber. Cut into thick rings.

3. Rub salt and caltrop starch on the pork tripe repeatedly. Rinse well. Boil some water and put in spring onion, ginger and rice wine. Blanch the pork tripe for 5 minutes. Cut into thick strips.

4. Clean the conches with a toothbrush. Remove the innards. Blanch in boiling water for 1 minute. Slice thick. Set aside. Blanch the Jinhua ham in boiling water to remove the stale oil on the surface.

5. Wipe the dried pork heel sinews with a dry cloth. Put them into cold oil in a wok. Then turn on low heat. Fry them for 30 minutes. Stir with wooden chopsticks from time to time. Cook until the sinews swell, bubble and turn deep golden. Drain. Soak them in water. Squeeze out excessive oil. Rinse well with water. Repeat the squeezing and rinsing for a few times. Then cut each sinew into halves.

6. Pour boiling water into a double-steamer. Put in sliced ginger, garlic, chicken, pork ribs, pork tripe, dried black mushrooms, dried scallops, ham and conches. Steam for 2 hours. Add pork heel sinews and sea cucumber. Steam for 30 more minutes. Season with rice wine at last. Serve.

肉燕 Meat
Skin
Dumplings

肉燕皮是福州特產，用
瘦肉及生粉剁成泥，再
壓成紙般薄片，曬乾而
成。燕皮包入餡料做成
燕皮餃，稱為「肉燕」，
吃起來爽口帶鮮味，與
一般的麵皮不同，是酒
席常用的湯餚。

【材料】
燕皮 20 張
上湯 5 杯

【餡料】
免治豬肉 100 克
蔥 1 條（切粒）
馬蹄 2 粒（切幼粒）
香菇 2 隻（切粒）
芫茜碎 2 湯匙
芹菜末少許
蔥酥及蒜酥各少許

【豬肉調味料】
鹽、胡椒粉及麻油各 1/4 茶匙
酒、生抽及粟粉各半茶匙

【湯調味料】
麻油及雞粉各 1 茶匙
鎮江黑醋 2 湯匙
喼汁 1 湯匙

【 Ingredients 】
20 sheets meat skin
5 cups premium stock

【 Filling 】
100 g ground pork
1 sprig spring onion (diced)
2 water chestnuts (diced finely)
2 dried black mushrooms (diced)
2 tbsps chopped coriander
finely chopped Chinese celery
deep-fried shallot
deep-fried garlic

【 Seasoning for ground pork 】

1/4 tsp salt
1/4 tsp ground white pepper
1/4 tsp sesame oil
1/2 tsp rice wine
1/2 tsp light soy sauce
1/2 tsp cornstarch

【 Seasoning for soup 】

1 tsp sesame oil
1 tsp chicken bouillon powder
2 tbsps Zhenjiang black vinegar
1 tbsp Worcestershire sauce

【做法】

1. 燕皮剪成 3 吋 x 5 吋小片，噴水令兩面濕透。

2. 豬肉及調味料拌勻，與其餘餡料混和，醃 30 分鐘。

3. 在燕皮上半部放上餡料半茶匙，包好，用手指抹少許水捏緊（以防煮時散開），
   成長尾燕餃。

4. 煮滾上湯，放入燕餃煮 6 至 7 分鐘，最後加湯調味料拌勻即可上桌。

【 小秘訣 • TIPS 】

- 肉燕的餡料多少，隨人喜歡，也可包成水餃狀，大小均佳，與湯麵、米粉湯或麵線湯同吃。

- 福州菜口味偏酸甜，故湯內加入黑醋。

- You may put more or less filling into the meat skin dumplings according to your taste. You may also wrap into regular dumpling shape in your preferred sizes. Alternatively, serve with noodle soup, rice vermicelli, or flour vermicelli.

- Fuzhou cuisine is characterized by a sweet and sour taste. Thus, I add Zhenjiang black vinegar to the soup.

【 Method 】

1. Cut the meat skin into rectangular pieces about 3" by 5". Spray water on both sides to wet them through.

2. Stir the seasoning into the ground pork. Mix well. Add other filling ingredients. Mix well. Leave it for 30 minutes.

3. On the upper half of each piece of meat skin, put 1/2 tsp of filling. Fold the skin downward. Wet your finger and secure the seam (so that the dumplings won't fall apart) into dumplings with a long tail.

4. Boil the stock. Cook the dumplings in the stock for 6 to 7 minutes. Add seasoning for the soup at last. Stir well and serve.

蜊仔湯 Baby Oyster Soup

【材料】

蚵仔 350 克
油條 1 條
番薯粉 2 湯匙
薑 2 大片（切絲）
乾蔥頭 2 粒（切片）
芫茜碎及蔥粒各 2 湯匙
油 2 湯匙
清雞湯 2 杯
水 1 杯

【蚵仔調味料】

酒及胡椒粉各 1 茶匙

【湯調味料】

紹酒 2 湯匙
老抽及生抽各 1 湯匙

【後下調味料】

麻油 1 湯匙
胡椒粉 1 茶匙

【 Ingredients 】

350 g baby oysters
1 deep-fried dough stick
2 tbsps sweet potato starch
2 large slices of ginger (finely shredded)
2 cloves shallot (sliced)
2 tbsps chopped coriander
2 tbsps diced spring onion
2 tbsps oil
2 cups chicken stock
1 cup water

【 Seasoning for baby oysters 】

1 tsp rice wine
1 tsp ground white pepper

【 Seasoning for the soup 】

2 tbsps Shaoxing wine
1 tbsp dark soy sauce
1 tbsp light soy sauce

【 Additional seasoning 】

1 tbsp sesame oil
1 tsp ground white pepper

【 做法 】

1. 油條切片，用油炸至香酥放在湯碗底。

2. 蚵仔的清洗方法，請參考第 56 頁。蚵仔瀝乾水分，加調味料及番薯粉輕輕拌勻，備用。

3. 鑊內燒熱油，爆香乾蔥頭及薑絲，澆入紹酒，傾入清雞湯及水煮滾，下湯調味料後加入蚵仔再煮滾，即可傾入湯碗，最後灑上蔥粒、芫茜碎、胡椒粉及麻油，即可上桌。

【 Method 】

1. Slice the deep-fried dough stick. Deep fry it in hot oil until crispy. Arrange on the bottom of a large soup bowl.

2. Rinse and clean the baby oysters according to the method on p.56. Drain well. Add seasoning and sweet potato starch. Stir softly. Set aside.

3. Heat oil in a wok. Stir fry shallot and ginger until fragrant. Sizzle with Shaoxing wine. Add chicken stock and water. Bring to the boil. Add seasoning for the soup. Put in the baby oysters. Bring to the boil and pour in the soup bowl over the deep-fried dough stick. Top with diced spring onion, chopped coriander, ground white pepper and sesame oil. Serve.

鮮雞精 Homemade Chicken Essence

這是一道老幼皆宜的補品。新鮮雞精鮮味濃郁，而且補身，與市面售賣的雞精不可同日而語，身體虛弱或婦女坐月均很適用。這麼有益、這麼簡單易做，怎能不試？

【材料】
　老雞 1 隻
　薑 1 片

【工具】
　有孔銻碗 1 個
　燉盅 1 個

【 Ingredients 】
　1 mature chicken
　1 slice ginger

【 Tools 】
　1 perforated metal bowl
　1 double-steamer

【做法】
1. 雞洗淨，去皮、去大骨及去油脂，用刀剁幼。
2. 雞肉用溫水半碗及鹽 1/4 茶匙拌勻，排於有孔銻碗內。
3. 薑片放在燉盅底，放上有孔銻碗，碗底到燉盅底需有足夠空間容納 2 碗水。
4. 水滾後計，蒸 2 小時，分泌出鮮雞汁，約可得 1 飯碗。

【 Method 】
1. Rinse and skin the chicken. Remove the big bones and trim off the fat. Then finely chop it with a knife.
2. Mix the chicken meat with 1/2 bowl of warm water and 1/4 tsp of salt. Mix well. Arrange on the perforated bowl.
3. Place the ginger on the bottom of a double-steamer. Place the perforated metal bowl on top. Beneath the metal bowl, there should be enough room in the double-steamer for two bowls of liquid.
4. Bring water to the boil. Steam for 2 hours. The essence of the chicken will start dripping into the double-steamer. You should get about 1 bowl at last.

【小秘訣 • TIPS】
- 視乎各人需要，可在燉盅底放數片當歸、洋參、花旗參、高麗參或杞子等。
- 感冒者及燥熱者不宜飲用。
- 想簡單方便的，可用雞胸肉或牛肉代替。
- Depending on your bodily condition, you may put a few slices of Dang Gui, American ginseng, Korean ginseng or a few Qi Zi in the double-steamer.
- Those suffering from influenza or Dryness-Heat should not consume.
- To simplify the step, use chicken breast or beef instead.

福建・清粥小菜
CONGEES AND STIR-FRIED

綿稠的稀飯、平實的佐粥小菜，

每道簡單的味道，

吃起來令人愉悦歡快！

Dense fluffy congee

simple homely stir-fries

straight forward taste

for straight forward pleasure

清淡暖胃之味！
The light and heart-warming flavours!

芋 Taro
粥 Congee

【小秘訣 • TIPS】

- 一般煲粥的話，1杯米（量米杯）會用6杯水（量米杯）；1 1/2杯米即用9杯水，因加入芋頭熬煮，故多加6杯水烹調。以上只是煲粥的烹調指引，可依自己喜愛的濃稠度，調整水的份量。

- 福建粥和廣東粥有很大分別，廣東粥綿滑；福建粥煮好後還清楚看見米粒。

- Generally speaking, when you make congee, you use 1 part rice to 6 parts water. That means we should add 9 cups of water to 1 1/2 cups of rice. However, this is no ordinary congee. The taro is very starch and will thicken the congee considerably. That's why we used 15 cups of water instead. Of course, the recipe is just a rough guideline. You can adjust the amount of water according to your preference.

- Fujian congee is quite different from Cantonese congee – the rice is mushy and almost completely melted in the latter; but you can see whole grains of rice in the former.

【 材 料 】
　芋頭 300 至 400 克（切塊）
　蚵仔乾 1 杯
　瑤柱碎 4 湯匙
　排骨或豬軟骨半斤
　煲粥米（或日本米）1 1/2 杯（量米杯）
　水 15 杯（量米杯）

【 調 味 料 】
　鹽、胡椒粉及麻油各 1 茶匙

【 Ingredients 】
　300 g to 400 g taro (cut into pieces)
　1 cup dried baby oysters
　4 tbsps dried scallops (not necessarily in whole)
　300 g pork ribs or pork gristles
　1 1/2 cups congee rice or Japanese pearl rice
　(measured with the cup that comes with the rice cooker)
　15 cups water
　(measured with the cup that comes with the rice cooker)

【 Seasoning 】
　1 tsp salt
　1 tsp ground white pepper
　1 tsp sesame oil

【 做 法 】
1. 瑤柱洗淨，去硬枕；蚵仔乾浸水，洗淨，去清蠔殼屑。

2. 排骨用粗鹽 1 大湯匙醃一晚或最少 2 小時，飛水備用。

3. 鍋內注入水及米，用大火煲滾，再轉中火煮 10 分鐘，放入其他材料煮滾，再轉小火煮 30 分鐘，最後加調味料拌勻，即可品嘗美味鹹粥。

【 Method 】
1. Rinse the dried scallops. Remove the tough tendon. Set aside. Soak the dried baby oysters in water until soft. Rinse and remove any broken shells.

2. Rub 1 large tbsp of coarse salt on the pork ribs. Leave them for at least 2 hours (preferably overnight). Rinse well. Blanch in boiling water. Drain.

3. In a pot, put in water and rice. Bring to the boil over high heat. Turn to medium heat and cook for 10 minutes. Put in all remaining ingredients and bring to the boil again. Turn to low heat. Simmer for 30 minutes. Season and serve.

番薯粥 Sweet Potato Congee

【材料】

番薯 1 個（約 300 克）
米 1 杯（量米杯）
水 8 至 10 杯（量米杯）

【Ingredients】

1 sweet potato (about 300 g)
1 cup rice (measured with the cup that comes with the rice cooker)
8 to 10 cups water (measured with the cup that comes with the rice cooker)

【做 法】

1. 番薯洗淨，去皮，切大塊，浸於水中備用。

2. 鍋內注入冷水，加入米用高火煮滾，不用加蓋，水滾後轉中火煮 20 分鐘，加入番薯煮 20 分鐘，關火，加蓋焗 10 分鐘，盛於小碗供吃。

【Method】

1. Rinse and peel the sweet potato. Cut into large chunks. Soak them in water for later use.

2. Pour cold water in a pot. Put in the rice and bring to the boil over high heat with the lid open. Turn to medium heat and cook for 20 minutes. Put in the sweet potato and cook for 20 minutes. Turn off the heat. Cover the lid. Leave it for 10 minutes. Serve in small bowls.

福建家庭常吃番薯粥代替米飯，也多作早餐或宵夜。若應酬繁多油膩吃不消，吃番薯粥配小菜，可清理腸胃。

【小秘訣 • TIPS】

• 準備下鍋煲粥時，才將番薯去皮，以免氧化變黑。

• 粥煮好後需要加蓋待一會，粥才會黏稠。

• Peel the sweet potato right before you put them into the congee, so that it won't be oxidized and turn black.

• After the congee is done, cover the lid and leave it for a while. The congee will be stickier and thicker this way.

菜脯蛋 Stir Fried
Dried Radish
and Egg

【材料】

菜脯 3/4 杯（切碎）

肉碎 1/4 杯

雞蛋 4 隻

乾葱茸 2 湯匙

蒜茸 1 湯匙

油 3 湯匙

【雞蛋調味料】

生抽、雞粉及粟粉各半茶匙

【肉碎調味料】

生抽及酒各 1/4 茶匙

【菜脯調味料】

糖 1 茶匙

麻油 1 茶匙

胡椒粉 1/4 茶匙

【 Ingredients 】

3/4 cup dried radish (finely chopped)

1/4 cup ground pork

4 eggs

2 tbsps chopped shallot

1 tbsp grated garlic

3 tbsps oil

【 Seasoning for the eggs 】

1/2 tsp light soy sauce

1/2 tsp chicken bouillon powder

1/2 tsp cornstarch

【 Seasoning for the ground pork 】

1/4 tsp light soy sauce

1/4 tsp rice wine

【 Seasoning for the dried radish 】

1 tsp sugar

1 tsp sesame oil

1/4 tsp ground white pepper

【 做 法 】

1. 雞蛋與調味料拂打成蛋液。

2. 肉碎與調味料炒香。

3. 菜脯洗淨，抹乾，剁碎。

4. 熱鑊下油，爆香乾葱茸，下菜脯炒至乾透及散發香味，加入蒜茸及菜脯調味料再炒，下肉碎拌炒，最後傾入蛋液炒香即成。

【 Method 】

1. Whisk the eggs with the seasoning.

2. Stir fry the ground pork and break it into tiny bits. Add seasoning and stir well.

3. Rinse the dried radish. Wipe dry and chop finely.

4. Heat oil in a wok and stir fry shallot until fragrant. Put in the dried radish. Stir until fragrant and heated through. Add grated garlic and seasoning for the dried radish. Stir again. Put in the ground pork and mix well. Pour in the whisked eggs from step 1 at last. Fry until fragrant and serve.

炒蒜仔 Stir Fried
Chinese Leeks
with Pork and
Dried Beancurd

【材料】
唐芹 50 克（切小段）
蒜仔（青蒜苗）110 克
肉絲 120 克
豆乾 1 塊（約 120 克）
榨菜 120 克
乾葱頭 2 湯匙（切片）
紅辣椒半隻（切絲）
油 4 湯匙

【醃料】
酒及生抽各 1 湯匙
麻油、糖及生粉各 1 茶匙

【調味料】
生抽、蠔油及酒各 1 湯匙
糖及鹽各半茶匙

【 Ingredients 】
50 g Chinese celery (cut into short lengths)
110 g Chinese leeks
120 g shredded pork
1 cube dried beancurd (about 120 g)
120 g Zha Cai (spicy pickled mustard tuber)
2 tbsps shallot (sliced)
1/2 red chilli (shredded)
4 tbsps oil

【 Marinade 】
1 tbsp rice wine
1 tbsp light soy sauce
1 tsp sesame oil
1 tsp sugar
1 tsp caltrop starch

【 Seasoning 】
1 tbsp light soy sauce
1 tbsp oyster sauce
1 tbsp rice wine
1/2 tsp sugar
1/2 tsp salt

【做法】
1. 蒜仔洗淨，切去末段約 1/3 部分，其餘切斜片。

2. 榨菜洗淨，切片後再切成絲，浸水 30 分鐘，瀝乾水分。

3. 肉絲用醃料拌勻，備用。豆乾洗淨，抹乾水分，片成薄片，再切絲。

4. 燒熱油 2 湯匙，放入豆乾炒至金黃色，下榨菜絲再炒香，盛起備用。

5. 熱鑊下油 3 湯匙，爆香乾葱頭及肉絲，加入唐芹及蒜仔炒至香味散發，再加入榨菜絲、豆乾絲、紅椒絲及調味料拌炒即成。

【 Method 】
1. Rinse the Chinese leeks. Cut off one-third of their root ends. Slice diagonally.

2. Rinse the Zha Cai. Slice and then shred. Soak it in water for 30 minutes. Rinse and drain.

3. Add marinade to the shredded pork. Mix well. Rinse the dried beancurd and wipe dry. Slice thinly and then shred it.

5. Heat 2 tbsps of oil in a wok. Stir fry the dried beancurd until golden. Add Zha Cai and stir until fragrant. Set aside.

6. Heat 3 tbsps of oil in a wok. Stir fry shallot and shredded pork until fragrant. Add Chinese celery and leeks. Stir until fragrant. Put in the fried Zha Cai and dried beancurd. Add red chilli and seasoning. Stir well. Serve.

花生辣豆乾

Spicy
Dried
Beancurd
with Peanuts

【材料】

五香豆乾 4 件
蒜茸及薑米各 1 湯匙
豆豉 2 湯匙
麻油 3 湯匙
紅辣椒 1 隻
炸花生 1 杯（做法參考第 61 頁）

【調味料】

胡椒粉半茶匙
生抽及辣椒油各 1 茶匙
糖 2 茶匙

【 Ingredients 】

4 cubes five-spice dried beancurd
1 tbsp grated garlic
1 tbsp finely diced ginger
2 tbsps fermented black beans
3 tbsps sesame oil
1 red chilli
1 cup deep-fried peanuts (please refer to p.61 for method)

【 Seasoning 】

1/2 tsp ground white pepper
1 tsp light soy sauce
1 tsp chilli oil
2 tsps sugar

【做法】

1. 豆乾洗淨，抹乾水分，切丁；紅辣椒切粒。

2. 燒熱麻油 3 湯匙，加入豆乾及調味料炒香，取出備用。

3. 原鑊爆香薑米、蒜茸、豆豉及紅辣椒，關火，豆乾回鑊拌勻，待涼，加炸花生即可。

【 Method 】

1. Rinse the dried beancurd. Wipe dry and dice it. Set aside. Dice the red chilli.

2. Heat 3 tbsps of sesame oil in a wok. Stir fry the dried beancurd and add seasoning. Set aside.

3. In the same wok, stir fry ginger, garlic, fermented black beans and red chilli in oil until fragrant. Turn off the heat. Put the dried beancurd back in. Toss well. Leave it to cool. Sprinkle deep-fried peanuts on top. Serve.

【 小 秘 訣 • TIPS】

嗜辣者可加指天椒 4 隻，特別惹味好吃！

For spicy food lovers, feel free to add 4 bird eye chillies to the above recipe. They will give the extra kick you need.

丁香魚仔炒肉絲

Stir Fried
Shredded Pork with
Dried Anchovy

【小秘訣 • TIPS】

炒肉絲時，鑊必須燒得很熱才下油，肉絲才不會黏底，而且鑊氣十足，所謂的鑊氣即猛鑊之氣，當菜餚起鑊前潷少許酒，能增加鑊氣及加強香味。

Before you put oil in the wok, make sure it is smoking hot. Then the shredded pork won't stick to the wok as much and add Wok Qi to the dish. In Chinese stir-fries, Wok Qi is an important concept – it is the searing and charring of food accompanied by caramelization when it is tossed quickly over very high heat. By adding a splash of wine at last also adds aroma and Wok Qi.

很多福建同鄉在家裏也
會品嘗這個家常小菜，
你也來試試吧！

【材料】
半肥瘦豬肉 150 克
罐裝豆豉 2 湯匙
丁香鹹魚仔半杯
乾葱頭 4 粒（切片）
油 2 湯匙

【醃料】
粟粉、雞粉、胡椒粉
及糖各半茶匙
生抽半湯匙

【後下材料】
葱 3 條（切粒）
蒜頭 3 粒（剁茸）
紅辣椒 1 隻（切幼圈）
麻油 1 湯匙
酒 2 湯匙（最後加入）

【 Ingredients 】
150 g half fatty pork
2 tbsps canned fermented black beans
1/2 cup dried anchovies
4 cloves shallot (sliced)
2 tbsps oil

【 Marinade 】
1/2 tsp cornstarch
1/2 tsp chicken bouillon powder
1/2 tsp ground white pepper
1/2 tsp sugar
1/2 tbsp light soy sauce

【 Garnishes 】
3 sprigs spring onion (diced)
3 cloves garlic (grated)
1 red chilli (cut into thin rings)
1 tbsp sesame oil
2 tbsps rice wine (added at last)

【做法】

1. 豬肉切絲，加醃料醃 30 分鐘，用油炒熟備用。

2. 鹹魚仔洗淨，瀝乾水分。

3. 燒熱油 2 湯匙，下乾葱頭爆至微黃，放入鹹魚仔炒至乾身，下豆豉及豬肉絲炒香，傾入後下材料拌勻，沿鑊邊澆入酒即成。

【 Method 】

1. Shred the pork. Add marinade and leave it for 30 minutes. Stir fry in a little oil until done. Set aside.

2. Rinse the anchovies. Drain well.

3. Heat 2 tbsps of oil in a wok. Stir fry shallot until lightly browned. Add anchovies and stir until dry. Add fermented black beans and shredded pork. Stir well. Put in the garnishes (except the wine). Then sizzle with rice wine along the rim of the wok. Serve.

醋浸三味 Trio Pickles

【材料】

子薑 450 克
榨菜 300 克
小蝦米 150 克
蒜茸 3 湯匙

【調味料】

美國蘋果醋 2 杯
糖 1 杯
檸檬 1 個（榨汁）

【Ingredients】

450 g young ginger
300 g Zha Cai (spicy pickled mustard tuber)
150 g dried small shrimps
3 tbsps grated garlic

【Seasoning】

2 cups America apple vinegar
1 cup sugar
1 lemon (squeezed)

【做法】

1. 子薑及榨菜洗淨，切小丁，用冷開水沖洗一次，瀝乾水分，用廚房紙吸乾水分。

2. 小蝦米洗淨，剪去黑色腸，用冷開水浸 5 分鐘，取出，用廚房紙吸乾水分。

3. 將子薑、榨菜、蝦米、蒜茸及調味料拌勻，放入玻璃瓶內，置於冰箱泡浸兩日，即可食用。

4. 進食時，用潔淨的餐具取吃，密封蓋，放在冰箱可儲存一個月。

【Method】

1. Rinse the young ginger and Zha Cai. Dice them. Rinse them with cold drinking water once more. Drain. Wipe dry.

2. Rinse the dried shrimps. Devein with a pair of scissors. Soak them in cold drinking water for 5 minutes. Drain. Wipe dry.

3. Mix together the young ginger, Zha Cai, dried shrimps, garlic and seasoning. Transfer into an air-tight glass container. Leave them in the fridge for 2 days. Serve.

4. When you serve, take the pickles out with clean utensils. Seal the lid well afterwards. They last well in the fridge for 1 month.

【小秘訣 • TIPS】

- 用冷開水沖洗子薑、榨菜及小蝦米，避免使用水喉水，令保存時間較長。

- 吃蒜茸不但對身體有益，亦增加香氣、殺菌及保鮮。

- I suggest rinsing the ingredients with cold drinking water instead of tap water because the pickles last longer this way.

- Raw garlic is not only good for health, but also adds aroma to the food, kills germs and preserves the food better.

福建 ● 小吃麵飯
FUJIAN SNACKS AND STAPLES

愛…鄉間的平淡滋味，

嘗一口家鄉菜包、鹹點，

感受故鄉純樸的情懷！

Love… the light taste of countryside food

A bite of rustic dumplings and savoury treats

A slice of unadorned life in my hometown

家鄉菜包

Rustic Steamed Dumplings with Shredded Radish Filling

福建菜包（蘿蔔絲包）是我家的常備點心，也是大人小孩的下午茶點，無論蒸或煎吃，都非常美味！

我經常會多弄一些菜包，兒子、媳婦及孫兒，甚至他們的朋友也很喜歡吃，這種家鄉風味的包子，外面食肆不是經常可以吃到的！

【皮料】
糯米粉 1 包（600 克）
粘米粉半包（300 克）
糖 1 湯匙
鹽 1 茶匙
溫水約 5 杯

【餡料】
曬乾蘿蔔絲 200 克
（北角春秧街街市有售）
蝦米粒 3/4 杯
香菇粒 3/4 杯
半肥瘦义燒 300 克（半斤，切粒）
乾葱頭 10 粒（切片，分兩次用）
粟粉 3 湯匙（後下）

【調味料 1】
胡椒粉 1 茶匙
生抽 1 湯匙
酒 1 湯匙

【調味料 2】
蠔油及水各 4 湯匙
麻油 2 湯匙
生抽 2 湯匙
雞粉及胡椒粉各 1 茶匙

【 Dumpling skin 】
600 g glutinous rice flour
300 g rice flour
1 tbsp sugar
1 tsp salt
5 cups warm water

【 Filling 】
200 g shredded dried radish (available at
Chun Yeung Street Market, North Point)
3/4 cup diced dried shrimps
3/4 cup diced dried black mushrooms
300 g diced half-fatty barbecue pork
10 cloves shallot (sliced and divided into
two equal portions)
3 tbsps cornstarch (added at last)

【 Seasoning 1】
1 tsp ground white pepper
1 tbsp light soy sauce
1 tbsp rice wine

【 Seasoning 2】
4 tbsps oyster sauce
4 tbsps water
2 tbsps sesame oil
2 tbsps light soy sauce
1 tsp chicken bouillon powder
1 tsp ground white pepper

【包法】

1. 取一塊粉糰做成窩形，放入多些餡料壓實，包好，皮薄餡多最佳，完成後掃上麻油及灑上芝麻。

2. 蒸籠內鋪上牛油紙，用叉子戳數次以疏氣，排上菜包（別太密）。

3. 燒滾水，加蓋，用中小火蒸20分鐘。水必須滾着，但別用猛火，才能保持菜包的形狀。

4. 吃不完的菜包可放入冰箱，享用時以小火煎透或用微波爐翻熱。

【To assemble】

1. Pinch one piece of dough into a pouch. Stuff it with some fillings. Seal the seam by pinching well. Roll it round and brush on sesame oil. Sprinkle with sesames.

2. Line a steamer with baking paper. Prick with a fork a few times to facilitate the circulation of steam. Arrange the dumplings on the steamer. Leave some space between them.

3. Boil the water and cover the lid. Steam the dumplings over medium-low heat for 20 minutes. The water must be boiling throughout the steaming process. Do not steam them over high heat. Otherwise, the dumplings may deform and burst. Serve.

4. Leftover dumplings can be stored in the fridge. Just pan-fry them or reheat them in a microwave oven before serving.

【小秘訣 • TIPS】

- 搓粉糰時，別一次過加入水，最好留下小半杯作為調整，因麵粉的乾濕程度有異。

- 弄包皮時，先煮熟少許粉，才與乾粉拌勻，搓出來的包皮才不會爆裂。

- 預備餡料時，最後加入粟粉，令餡料帶黏性，容易包裹。

- 餡料必須涼透才包，外皮才不容易破裂。

- When you add water to the dry ingredients to make the dough, do not put in all the water all at once. Add slightly more than 1/2 a cup first and add the rest a little at a time if needed. The amount of water added depends on the dryness of the flour and atmospheric humidity. If you pour in 1 cup all at once, the dough can be too wet.

- I used a tempering technique to make the dough. By mixing in cooked dough into the dry ingredients, the dumpling skin becomes more resilient and less likely to burst.

- I added some cornstarch to the filling to bind the ingredients together. It makes the filling less watery and easier to handle.

- Wait until the filling is completed cold before stuffing the dough with it. Otherwise, the skin may crack easily.

五香牛脹

Five-Spice
Marinated Beef Shin

牛脹用途多多，加生菜番茄做成三文治、沙律，也可配湯麵、飯或粥，也是下酒的冷盤好菜。

【材料】
　牛䐁 3 件（約 750 克）
　洋蔥 250 克
　油半杯
　蔥 6 條（切段）
　乾蔥頭、蒜頭各 6 粒（拍扁）
　薑 3 厚片

【醃料】
　紹酒 1 湯匙
　粗鹽 1 湯匙
　五香粉 2 茶匙

【滷水汁料】
　紹酒半杯
　蠔油 1/3 杯
　柱侯醬 2 湯匙
　生抽 2 湯匙
　冰糖碎 2 湯匙
　粗粒黑胡椒 1 湯匙
　鹽 1 茶匙
　八角 4 粒
　花椒粒 1 湯匙
　滾水 2 杯

【滷水蘸汁】
　蒜茸 1 湯匙
　老抽 2 湯匙
　糖、麻油、辣椒油、雞湯、白醋、
　芫茜末或蔥末各 1 湯匙

【 Ingredients 】
　3 pieces of beef shin (about 750 g)
　250 g onion
　1/2 cup oil
　6 sprigs spring onion (cut into short lengths)
　6 cloves shallot (gently crushed)
　6 cloves garlic (gently crushed)
　3 thick slices ginger

【 Seasoning 】
　1 tbsp Shaoxing wine
　1 tbsp coarse salt
　2 tsps five-spice powder

【 Spiced marinade 】
　1/2 cup Shaoxing wine
　1/3 cup oyster sauce
　2 tbsps Chu Hou sauce
　2 tbsps light soy sauce
　2 tbsps crushed rock sugar
　1 tbsp coarsely ground black pepper
　1 tsp salt
　4 cloves star anise
　1 tbsp Sichuan peppercorns
　2 cups boiling water

【 Dipping sauce 】
　1 tbsp grated garlic
　2 tbsps dark soy sauce
　1 tbsp sugar
　1 tbsp sesame oil
　1 tbsp chilli oil
　1 tbsp chicken stock
　1 tbsp white vinegar
　1 tbsp grated coriander or spring onion

【做法】

1. 牛脹用醃料醃 4 小時，醃一晚最佳。

2. 瓦煲內燒熱油，爆香薑、乾葱頭、蒜頭及洋葱，放入牛脹爆香，葱段排於煲底，傾入滷水汁料煮滾，以中火煮 3 分鐘，轉小火燜煮（小牛脹燜 75 分鐘；大牛脹燜約 90 分鐘）。

3. 牛脹盛起，待涼，放入冰箱冷藏至硬身（容易切成薄片）。

4. 享用時切薄片，澆上滷水汁供食，或以滷水蘸汁伴吃。

【 Method 】

1. Add seasoning to the beef shin and mix well. Leave it for at least 4 hours (or preferably overnight).

2. In a clay pot, heat some oil. Stir fry ginger, shallot, garlic and onion until fragrant. Sear the beef shin until lightly browned. Arrange the spring onion on the bottom of the clay pot. Pour in the spiced marinade ingredients. Bring to the boil. Turn to medium heat and cook for 3 minutes. Turn to low heat and simmer until done. Small pieces of beef shin take 75 minutes to cook. Big pieces need 90 minutes.

3. Take the beef shin out of the marinade. Leave it to cool. Then refrigerate until stiff so that you can slice it more easily.

4. Before you serve, slice the beef shin thinly. Drizzle with some of the spiced marinade. Or serve with the dipping sauce on the side.

【小秘訣 • TIPS】

• 滷牛脹前先用油半煎炸，可去除牛脹的騷味，而且縮短燜煮時間。

• 細條的牛脹又名花脹，口感較佳。

• 以上的滷水汁料可滷製牛肚，將整塊急凍牛肚飛水，放入滷水汁內煮 1 至 1 1/2 小時，切長條供食。

• 不吃牛肉者，可用豬脹代替。

• Deep-frying the beef shin before marinating it helps remove the musky taste of the beef and shortens the marinating time.

• Smaller beef shins have more connective tissues interlaced between the muscles. They actually taste better than the large ones.

• The spiced marinade in this recipe also works for beef tripe. Just thaw the beef tripe and blanch it in boiling water. Then cook it in the marinade for 1 to 1 1/2 hours. Cut into long strips and serve.

• Those who don't eat beef may use pork shin instead.

芋
圓

Deep Fried
Taro
Dumplings

【 Ingredients 】
600 g taro
1/2 cup caltrop starch

【 Filling 】
200 g pork (finely chopped)
3 tbsps diced dried
black mushrooms
2 tbsps chopped dried shrimps
1 cup finely chopped
spring onion
2 tbsps sliced shallot
1/2 tbsp grated garlic

【 Seasoning 】
1/2 tbsp salt
1/2 tbsp five-spice powder
2 tbsps sugar
4 tbsps caltrop starch
2 tbsps oil
water

【 Marinade 】
1 tsp salt
1 tsp ground white pepper
1 tsp sugar
1 tsp five-spice powder
1 tbsp light soy sauce
1 tbsp oyster sauce
1 tbsp rice wine
1 tbsp cornstarch

【材料】
芋頭 600 克
生粉半杯

【餡料】
豬肉 200 克（剁碎）
香菇 3 湯匙（切小粒）
蝦米碎 2 湯匙
蔥末 1 杯
乾蔥片 2 湯匙
蒜茸半湯匙

【調味料】
鹽及五香粉各半湯匙
糖 2 湯匙
生粉 4 湯匙
油 2 湯匙
水適量

【醃料】
鹽、胡椒粉、糖及五香
粉各 1 茶匙
生抽、蠔油、酒及粟粉
各 1 湯匙

【小秘訣 • TIPS】

- 芋圓要分批下油鍋炸，因家中的用油量不及酒樓般。

- 若芋圓在滾油內炸太久，芋圓容易變形或爆裂。

- Deep fry the taro dumplings in a few batches because household woks don't hold as much oil as those in restaurant kitchens.

- Do not fry the taro dumplings for too long. Or else they would deform or burst.

【做法】

1. 芋頭去皮、切片，蒸 25 分鐘至軟透，取出，趁熱壓成泥，加調味料搓勻（若太乾，可加少許水）。

2. 豬肉與醃料拌勻；蝦米洗淨，略浸，剁碎。

3. 燒熱油下乾蔥片及蒜茸爆香，加入肉碎、蝦米及菇粒炒香，拌入蔥末，取出備用。

4. 取芋泥 38 克捏成窩形，放入餡料 1 茶匙，搓圓成球狀，共製成 24 至 26 個，撲上少許生粉，放入熱油內用中火炸至香酥約 2 至 3 分鐘，以五香椒鹽伴吃。

【Method】

1. Peel and slice the taro. Steam for 25 minutes until tender. Mash while still hot. Add seasoning and mix well. The mashed taro should bind itself well. If it seems too dry, add a little water.

2. Add the marinade to the pork. Stir well. Set aside. Rinse the dried shrimps. Soak them in water briefly. Drain and finely chop them.

3. Heat oil in a wok. Stir fry shallot and garlic. Add ground pork, dried shrimps and dried black mushrooms. Stir until fragrant. Stir in spring onion. This is the filling.

4. Take 38 g of mashed taro. Roll into a ball. Dig a hole at the centre and stuff in 1 tsp of filling. Seal the seam and roll into a ball again. The recipe here should make 24 to 26 dumplings. Lightly coat them in caltrop starch. Deep fry in oil over medium heat for about 2 to 3 minutes until golden. Drain and serve with five-spice peppered salt on the side as a dip.

這是閩南食品，作為宴客或家中茶點均非常適宜，越吃越喜愛，而且製法容易，上桌時很有特色。冬天吃熱呼呼的；夏天可放入冰箱冷吃，冰冰涼涼的十分可口，故在台灣及南洋一帶很受歡迎！

此食譜是最基本的做法，質感軟滑，粉漿也可加入番薯粉或糯米粉，餡料可隨個人喜好而更改，各適其適。

碗糕

Steamed
Rice Cakes in Bowls

【材料】
粘米粉 3 杯
澄麵 3 湯匙
鹹蛋黃 6 個
冷水 3 杯
滾水 6 杯
芫茜碎或葱粒半杯

【餡料】
肉碎 150 克
香菇 4 隻
蝦米 2 湯匙
甜菜脯 2 湯匙
乾葱頭 4 粒（切碎）

【調味料】
生抽 1 湯匙
蠔油 1 湯匙
胡椒粉 1/4 茶匙
雞粉 1/4 茶匙
鹽 1/4 茶匙
生粉 1 茶匙

【芡汁】
蒜頭 2 粒（剁茸）
生抽 2 湯匙
蠔油 2 湯匙
水 1 1/2 杯
麻油 2 茶匙
生粉水 2 茶匙

【 Ingredients 】
3 cups rice flour
3 tbsps wheat starch
6 salted egg yolks
3 cups cold water
6 cups boiling water
1/2 cup chopped coriander or diced spring onion

【 Filling 】
150 g ground pork
4 dried black mushrooms
2 tbsps dried shrimps
2 tbsps sweet dried radish
4 cloves shallot (finely chopped)

【 Seasoning 】
1 tbsp light soy sauce
1 tbsp oyster sauce
1/4 tsp ground white pepper
1/4 tsp chicken bouillon powder
1/4 tsp salt
1 tsp caltrop starch

【 Glaze 】
2 cloves garlic (grated)
2 tbsps light soy sauce
2 tbsps oyster sauce
1 1/2 cups water
2 tsps sesame oil
2 tsps caltrop starch solution

【 小秘訣 • TIPS 】

- 蒸米糕的時間可自行調整，視乎碗的大小而定。

- 用牙籤插入米糕，取出，不黏著牙籤的即是熟透。

- As the bowls you use might be of different sizes from mine, please adjust the steaming time on your own.

- To check the doneness of the rice cakes, insert a toothpick at the centre of the rice cake. If it comes out clean without any gooey mess sticking to it, the rice cake is done.

【 做法 】

1. 甜菜脯洗淨，剁碎；香菇及蝦米用水浸軟，洗淨，切小粒。

2. 鹹蛋黃切半，於刀背抹油，用刀背壓扁成薄片。

3. 鑊內下油爆香乾葱頭，加入肉碎炒熟，再下其餘餡料及調味料炒香，盛起備用。

4. 粘米粉及澄麵用冷水 3 杯調成粉漿。燒滾水 6 杯，慢慢傾入粉漿，傾入時不斷攪拌至糊狀，關火。

5. 小碗內抹上油，碗底放入一片鹹蛋黃，傾入米糊至 5 分滿，放入餡料，再加入米糊至滿，抹平，蒸 12 至 20 分鐘，取出。

6. 煮滾芡汁，米糕倒扣碟上，澆上芡汁，加上芫茜或葱粒，伴辣醬品嘗更美味，是朋友聚會或孩子下午茶點之美食。

【 Method 】

1. Rinse the dried radish and finely chop it. Set aside. Soak the black mushrooms and dried shrimps in water until soft. Rinse well and dice finely.

2. Cut the salted egg yolks into halves. Brush some oil on the flat side of a knife. Press each half of egg yolk into a flat patty with the knife.

3. Heat oil in a wok and stir fry shallot until fragrant. Add pork. Stir until done. Put in the rest of the filling and seasoning. Stir fry until fragrant. Set aside.

4. Put the rice flour and wheat starch into a mixing bowl. Add 3 cups of cold water. Mix well into a batter. Then boil 6 cups of water in a pot. Slowly pour the batter into the boiling water while stirring continuously. Cook until thick. Turn off the heat.

5. Grease 12 small bowls. Put a piece of salted egg yolk in each bowl. Fill them with the rice batter up to half full. Arrange some filling on top. Then fill the bowls with the remaining rice batter. Smooth the surface and steam for 12 to 20 minutes.

6. Cook the glaze in a wok. Turn the steamed rice cakes out on a plate. Drizzle with the glaze and garnish with coriander or spring onion. You may also serve with chilli sauce on the side. It makes a great treat for gatherings or an afternoon snack for children.

福建炒麵

Fujian
Chow Mein
~Fried Noodles~

【材料】
　油麵 600 克
　豬肉 200 克
　韭菜 50 克
　椰菜 150 克
　雞蛋 3 隻
　蒜頭及乾葱頭各 2 粒（切片）
　芫茜 2 棵
　雞湯 2 杯
　豬油 2 湯匙
　炸葱酥 2 湯匙
　油適量

【醃料】
　生抽 1 湯匙
　胡椒粉、糖及粟粉各 1 茶匙

【調味料】
　沙茶醬及生抽各 1 湯匙
　甜豉油 2 湯匙（做法見第 135 頁）
　紹酒及鎮江醋各 2 湯匙（後下）

【* 甜豉油材料】
　李錦記老抽 1 瓶
　砂糖 1 1/4 杯
　糯米粉 3 湯匙（與水 5 湯匙拌勻）

【 Ingredients 】
　600 g fresh thick egg noodles
　200 g pork
　50 g chives
　150 g white cabbage
　3 eggs
　2 cloves garlic (sliced)
　2 cloves shallot (sliced)
　2 sprigs coriander
　2 cups chicken stock
　2 tbsps lard
　2 tbsps deep-fried shallot
　oil

【 Marinade 】
　1 tbsp light soy sauce
　1 tsp ground white pepper
　1 tsp sugar
　1 tsp cornstarch

【 Seasoning 】
　1 tbsp Sa Cha sauce
　1 tbsp light soy sauce
　2 tbsps sweet soy sauce (refer to
　p.135 for method)
　2 tbsps Shaoxing wine (added at last)
　2 tbsps Zhenjiang black vinegar (added at last)

【*Ingredients of sweet soy sauce 】
　1 bottle Lee Kum Kee dark soy sauce
　1 1/4 cups sugar
　3 tbsps glutinous rice flour
　(mixed with 5 tbsps of water)

【甜豉油做法】

老抽及糖用小火熬煮 1 小時，煮時切勿攪動，最後加入糯米粉水勾芡即可。

【做法】

1. 豬肉切成絲，加醃料醃 30 分鐘，用油炒熟備用。

2. 韭菜切段，用油略炒；椰菜切幼絲，用油炒至軟身，備用。

3. 雞蛋拂打成蛋汁，煎成蛋皮，切粗絲。

4. 燒熱豬油爆香乾葱頭及蒜頭，下油麵及椰菜拌炒一會，加入調味料及雞湯再炒，加蓋燜煮 5 分鐘至油麵入味，下肉絲、蛋絲及韭菜拌勻，最後澆入紹酒及鎮江醋，灑上芫茜碎及炸葱酥即可上碟。

【Method of sweet soy sauce 】

Cook dark soy sauce and sugar over low heat for 1 hour. Do not stir while cooking. Stir in the glutinous rice flour solution at last.

【 Method 】

1. Shred the pork. Add marinade and mix well. Leave it for 30 minutes. Stir fry in a little oil until done. Set aside.

2. Cut the chives into short lengths. Stir fry briefly in some oil. Set aside. Finely shred the cabbage. Stir fry in oil until soft. Set aside.

3. Whisk the eggs and fry into an omelette. Shred coarsely.

4. Heat lard in the wok and stir fry shallot and garlic until fragrant. Put in the egg noodles and cabbage. Stir briefly. Add seasoning and chicken stock. Stir again. Cover the lid and cook for 5 minutes until the noodles pick up the seasoning. Add shredded pork, egg omelette and chives. Lastly sizzle with Shaoxing wine and Zhenjiang vinegar. Sprinkle with chopped coriander and deep-fried shallot. Serve.

【小秘訣 • TIPS】

- 甜豉油不自行調製的話，也可在台灣或東南亞食材店購買。康怡吉之島有售。

- 油麵燜至濕軟入味，味道才最佳。若想香味濃郁，可多加 2 湯匙豬油炒煮。

- To save time and effort, you can get sweet soy sauce in a bottle from Taiwanese or Southeast Asian deli stores. The Jusco in Kornhill also carries it.

- The egg noodles should be braised until soft and flavourful for the best result. If you prefer a stronger taste, add a couple more tablespoons of lard when you stir fry the noodles.

福建炒米粉　Fried Rice Vermicelli

【材料】
新竹幼米粉 200 克
腩肉（連皮）150 克
銀芽 140 克
韭菜 60 克
小蝦米 2 湯匙
香菇 4 隻
雞蛋 3 隻
乾葱頭 3 粒（切片）
豬油 3 湯匙
葱酥及蒜酥各 2 湯匙
芫茜 2 棵（切碎）

【雞蛋調味料】
鹽半茶匙

【調味料 1】
老抽、糖及鹽各 1 茶匙

【調味料 2（調勻）】
生抽 3 湯匙
黑醋、糖及麻油各 1 湯匙
胡椒粉半茶匙
清雞湯及冷開水各 1 杯

【 Ingredients 】

200 g fine rice vermicelli in Hsinchu style
150 g pork belly (with skin on)
140 g mung bean sprouts
60 g chives
2 tbsps dried small shrimps
4 dried black mushrooms
3 eggs
3 cloves shallot (sliced)
3 tbsps lard
2 tbsps deep-fried shallot
2 tbsps deep-fried garlic
2 sprigs coriander (finely chopped)

【 Seasoning for the eggs 】

1/2 tsp salt

【 Seasoning 1 】

1 tsp dark soy sauce
1 tsp sugar
1 tsp salt

【 Seasoning 2 (mixed well) 】

3 tbsps light soy sauce
1 tbsp Zhenjiang black vinegar
1 tbsp sugar
1 tbsp sesame oil
1/2 tsp ground white pepper
1 cup chicken stock
1 cup cold drinking water

【 小秘訣 • TIPS 】

建議用筷子炒米粉，將米粉挑鬆，令米粉炒得爽口入味，而且保持美觀賣相。

I prefer stir-frying rice vermicelli with a pair of chopsticks instead of a spatula. The rice vermicelli end up fluffier this way and they pick up the seasoning much better. The rice vermicelli can also be kept in long strands without being broken into bits and pieces, meaning better presentation too.

【 做法 】

1. 滾水內放入少許油及鹽，下銀芽及韭菜焓10秒，盛起。

2. 蝦米洗淨，用水浸20分鐘；香菇用水浸透，去蒂，切絲；雞蛋用鹽調味炒熟，切絲；腩肉連皮切成絲。

3. 米粉剪成長段，浸於滾水待3分鐘，撥散，瀝乾水分。

4. 燒熱油2湯匙，放入乾葱片爆至微黃，下腩肉煎香兩面，加入調味料（1）拌勻，下香菇及蝦米拌炒，傾入調味料（2）及米粉炒勻，待汁液收乾，下雞蛋、芽菜、韭菜、芫茜、葱蒜酥及豬油，炒勻即可上碟。

【 Method 】

1. Boil some water. Add a little oil and salt. Blanch the mung bean sprouts and chives for 10 seconds. Drain.

2. Rinse the dried shrimps. Soak them in water for 20 minutes. Drain and set aside. Soak the black mushrooms in water until soft. Cut off the stems and shred them. Set aside. Crack the eggs and season with 1/2 tsp of salt. Whisk well. Fry into an omelette. Shred it and set aside. Cut the pork belly into thin strips with the skin on.

3. Cut the rice vermicelli into long strands. Soak them in boiling hot water for 3 minutes before use. Scatter with a pair of chopsticks. Drain well.

4. Heat 2 tbsps of oil in a wok. Stir fry shallot until lightly browned. Put in the pork belly and sear until both sides golden. Add seasoning (1) and stir well. Put in black mushrooms and dried shrimps. Mix well. Then pour in seasoning (2) and the rice vermicelli. Stir well and cook until the sauce reduces. Add shredded egg omelette, bean sprouts, chives, coriander, lard, deep-fried garlic and shallot. Toss well and serve.

高麗菜飯

Cabbage
Rice

腩肉醃一晚，連皮炒，味道鹹香！另外，腩肉配豆豉韭菜炒煮，加糖調味，配飯、伴粥或佐酒，非常美味！工作忙碌沒時間弄菜做飯，煮一鍋高麗菜飯，一家大小也能吃得飽足！

【材 料】

　高麗菜（椰菜）600 克
　急凍腩肉（連皮）150 克
　蝦米 2 湯匙
　香菇 4 隻
　乾葱頭 3 粒（切片）
　米 3 杯（量米杯）
　冷水 2 1/2 杯（量米杯）
　油 2 湯匙

【醃 料】

　粗鹽半湯匙

【調味料 1】

　糖及鹽各 1 湯匙

【調味料 2】

　麻油 1 湯匙
　葱粒 2 湯匙

【做 法】

1. 蝦米及香菇用水浸軟，香菇剪去硬蒂，切粒。

2. 腩肉用醃料醃 2 小時，若醃一晚食味更佳，切粗條。

3. 椰菜切粗絲，加調味料（1）略炒，盡量別釋出水分。

4. 燒熱油爆香乾葱片至微黃，放入腩肉煎香，下蝦米及香菇炒香，傾入米及椰菜拌勻，轉放電飯煲內，注入冷水煮至米飯熟透，用筷子拂鬆，焗 10 分鐘，傾入調味料（2）拌勻，盛於碗內，上桌時配炸花生伴吃。

【 Ingredients 】

　600 g white cabbage (the flat variety if possible)
　150 g frozen pork belly (with skin on)
　2 tbsps dried shrimps
　4 dried black mushrooms
　3 cloves shallot (sliced)
　3 cups rice (measured with the cup that comes with the rice cooker)
　2 1/2 cups cold water (measured with the cup that comes with the rice cooker)
　2 tbsps oil

【 Marinade 】

　1/2 tbsp coarse salt

【 Seasoning 1】

　1 tbsp sugar
　1 tbsp salt

【 Seasoning 2】

　1 tbsp sesame oil
　2 tbsps diced spring onion

【 Method 】

1. Soak the dried shrimps and black mushrooms in water until soft. Cut off the stems of the black mushrooms and dice them.

2. Add marinade to the pork belly and mix well. Leave it for 2 hours (or preferably overnight for full flavour). Cut into thick strips.

3. Coarsely shred the cabbage. Stir fry the cabbage in oil with seasoning (1). Try to keep it as dry as possible. Set aside.

4. Heat oil in a wok. Stir fry the shallot until lightly browned. Then sear the pork belly. Add dried shrimps and black mushrooms. Stir fry until fragrant. Add rice and cabbage. Stir well. Transfer the mixture into a rice cooker. Add cold water and turn on the cooker. When the rice is done, fluff it up with a pair of chopsticks. Cover the lid again and leave it for 10 minutes. Add in seasoning (2). Stir well. Serve in bowls with deep-fried peanuts.

鹹肉糭 Pork Zongzi ~Glutinous Rice Dumplings Wrapped in Bamboo Leaves~

【材料】（約做 35 隻）
糯米 2 公斤
腩肉 2 斤
香菇 18 隻
栗子 35 粒
大蝦米或蝦乾 35 隻
浸香菇及蝦米水共 3 杯
糉葉 80 張

【爆香材料】
薑 3 厚片
蒜頭及乾葱頭各 8 粒
油 3 湯匙

【滷水材料】
八角 6 粒
月桂葉 6 片
桂皮 1 小塊
甘草 6 片
陳皮 1/4 角
白胡椒粒 1 湯匙
花椒粒 2 湯匙

【調味料】
酒半杯
蠔油半杯
生抽 1/4 杯
雞粉半湯匙
五香粉半湯匙
胡椒粉半湯匙
冰糖 50 克

【糯米調味料】
五香粉半湯匙
老抽 3 湯匙
滷水汁 1 1/2 杯

【 Ingredients 】(makes 35 dumplings)
2 kg glutinous rice
1.2 kg pork belly
18 dried black mushrooms
35 chestnuts
35 dried large prawns or dried shrimps
3 cups of soaking water (from soaking black mushrooms and dried shrimps)
80 bamboo leaves

【 Aromatics 】
3 thick slices ginger
8 cloves garlic
8 cloves shallot
3 tbsps oil

【 Spices 】
6 cloves star anise
6 dried bay leaves
1 small piece cassia bark
6 pieces liquorice
1/4 dried tangerine peel
1 tbsp white peppercorns
2 tbsps Sichuan peppercorns

【 Seasoning 】
1/2 cup rice wine
1/2 cup oyster sauce
1/4 cup light soy sauce
1/2 tbsp chicken bouillon powder
1/2 tbsp five-spice powder
1/2 tbsp ground white pepper
50 g rock sugar

【 Seasoning for glutinous rice 】
1/2 tbsp five-spice powder
3 tbsps dark soy sauce
1 1/2 cups spiced marinade

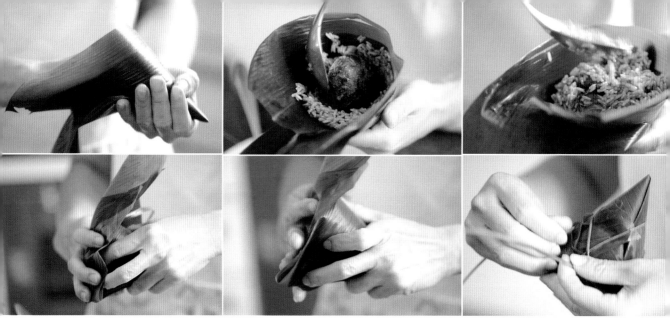

【 餡料做法 】

1. 糯米洗淨，用水浸 4 小時，
   瀝乾水分，備用。

2. 腩肉洗淨，抹乾水分，切成
   1 吋 x 1 1/2 吋大塊。

3. 香菇洗淨，用水浸軟，去硬
   蒂，開邊。浸香菇水留用。

4. 栗子放入滾水內煮 15 分鐘，
   取出，去皮。

5. 蝦米洗淨，用水浸 20 分鐘，
   取出備用。浸蝦米水留用。

6. 燒熱油，放入爆香材料用中
   火炒至微黃，待香氣四溢時
   加入滷水材料略炒，加入腩
   肉及香菇拌炒 5 分鐘，下調
   味料炒香，轉放瓦煲內，傾
   入浸香菇及蝦米水 3 杯，
   煮滾後轉小火燜煮 90 分鐘，
   取出腩肉及香菇備用。

7. 栗子及蝦米放於上述的滷水
   汁內煮 15 分鐘，盛起備用。

8. 鑊內燒熱油 3 湯匙，放入糯
   米及糯米調味料拌炒均勻，
   盛起備用。

【 To prepare the filling 】

1. Rinse the glutinous rice. Soak it water for 4 hours. Drain
   well.

2. Rinse the pork belly. Wipe dry. Cut into chunks about 1"
   by 1 1/2" big.

3. Rinse the black mushrooms. Soak them in water until
   soft. Cut off the stems and cut each into halves. Save the
   soaking water for later use.

4. Boil the chestnuts in water for 15 minutes. Drain. Shell
   and peel.

5. Rinse the dried shrimps. Soak them in water for 20
   minutes. Drain. Set the soaking water aside for later use.

6. Heat oil in a wok. Stir fry the aromatics over medium
   heat until lightly browned and fragrant. Put in the spices
   and stir briefly. Then put in the pork belly and black
   mushrooms. Stir for 5 minutes. Add seasoning. Stir until
   fragrant. Transfer into a clay pot. Pour in 3 cups of
   liquid (including the water from soaking dried shrimps
   and black mushrooms. Add water to make 3 cups if
   needed.). Bring to the boil over high heat. Then turn to
   low heat and simmer for 90 minutes. Set aside the pork
   and black mushrooms.

7. Put the chestnuts and dried shrimps into the same
   marinade from step 6. Cook for 15 minutes. Set aside.

8. Heat 3 tbsps of oil in a wok. Stir fry the glutinous rice
   and seasoning for glutinous rice until well incorporated.
   Set aside.

【小秘訣 • TIPS】
餡料可加入一整粒大
瑤柱或小鮑魚1隻，
滋味無窮！

For a luxurious variation,
add a big dried scallop
or a baby abalone to the
filling!

【包糉方法】

1. 每塊糉葉用刷清洗乾淨，放於鑊內用滾水煮滾，關火，加蓋，糉葉浸至軟身（約30分鐘），取出，抹乾水分。

2. 取糉葉兩張做成三角漏斗形，放入糯米1湯匙、栗子、蝦乾及腩肉各1件、香菇半隻，再排上糯米，蓋上糉葉，用水草綁緊，完成的糉有4隻角為標準。

3. 大湯煲內燒滾水，放入肉糉（水要蓋過面），用中大火煮1小時15分鐘即成（期間必須注入滾水，保持水浸過糉面）。

【 To assemble 】

1. Clean each bamboo leaf well with a brush. Rinse and place in a wok. Add water to cover. Bring to the boil. Turn off the heat and cover the lid. Leave the bamboo leaves in the hot water for 30 minutes. Drain. Wipe dry.

2. Fold two bamboo leaves together into an inverted pyramid shape. Put 1 tbsp of glutinous rice inside the leaves. Top with a chestnut, a dried shrimp, a piece of pork belly and half a black mushroom. Top with more glutinous rice. Fold the loose ends of the bamboo leaves to cover the filling well. Tie a reed string tightly around the bamboo leaves. The dumpling should have 4 pointy corners when done.

3. Boil water in a large stock pot. Put in the dumplings. There should be enough water to cover all dumplings. Cook over medium-high heat for 1 hour 15 minutes. Check the water level from time to time throughout the cooking process. Add boiling water if needed to keep the dumplings covered.

福建魚圓湯麵 Fujian Noodle Soup with Fish Balls

【材料】
　福建麵 2 個
　蜆仔 600 克
　連殼鮮蝦 300 克
　大白菜 600 克
　福建魚圓 300 克
　雞蛋 4 隻
　蒜仔 6 條
　芫茜碎及唐芹碎各適量
　番薯粉 1 杯

【調味料】
　鮮露 2 湯匙
　魚露 2 湯匙
　生抽 3 湯匙（可加減）
　麻油、胡椒粉及拔蘭地酒各適量（後下）

福建麵的特點是湯汁濃稠、麵條滑口，在東南亞一帶很受歡迎。湯麵也可加入魚頭或貝殼類海鮮等，材料多樣化，配上蜆仔及魚圓品嘗，是福建傳統的吃法。

【做法】
1. 鮮蝦連殼洗淨，去腸，剪去鬚腳；蜆仔用粟粉清洗，用手摸一遍沒有殼屑（清洗方法見第 56 頁）。

2. 大白菜及蒜仔切粗絲。雞蛋拂勻，加生抽炒碎及散發香味。

3. 蜆仔瀝乾水分，放在大碗內，與番薯粉輕輕拌勻，備用。

4. 福建麵剪成長段，放入大半鍋冷水內，用大火煮滾，下大白菜煮 10 分鐘，加入魚圓、蝦、雞蛋及蒜仔再煮 5 分鐘，下調味料拌勻，試味。最後加入蜆仔煮至麵湯大滾即可關火。

5. 吃麵時，加入芫茜、唐芹、麻油、胡椒粉及拔蘭地酒拌勻，鮮味可口。

【 Ingredients 】

2 bundles Fujian noodles
600 g baby oysters (shelled)
300 g prawns (with shells)
600 g Tianjian cabbage
300 g Fujian fish balls
4 eggs
6 stems Chinese leeks
chopped coriander
chopped Chinese celery
1 cup sweet potato starch

【 Seasoning 】

2 tbsps Maggi seasoning
2 tbsps fish gravy
3 tbsps light soy sauce (adjust the amount after tasting)
sesame oil, ground white pepper, brandy (added at last)

【 Method 】

1. Rinse the prawns. Devein and cut off the legs and antennae. Set aside. Rub cornstarch into the baby oysters and rinse well. Double check with your hands to see if there's any broken shell (refer to p.56 for method).

2. Cut Tianjin cabbage and Chinese leeks into thick strips. Set aside. Whisk the eggs. Add light soy sauce to the eggs and stir fry into an scramble egg.

3. Drain the baby oysters and transfer into a big bowl. Add sweet potato starch and mix gently.

4. Cut the noodles into long strands. Place them into a pot of cold water. Turn on high heat and bring to the boil. Put in the Tianjin cabbage and cook for 10 minutes. Add fish balls, prawns, scrambled egg and Chinese leeks. Cook for 5 more minutes. Add seasoning and mix well. Taste it and season accordingly. Put in the baby oysters at last and bring to a vigorous boil. Turn off the heat. Serve.

5. Right before serving, sprinkle with coriander, Chinese celery, sesame oil, ground white pepper and brandy and stir well.

【 小秘訣 • TIPS】

• 湯麵要煮得濃稠，美味好吃，緊記冷水下麵條！

• 福建魚圓即是有肉餡之福州魚蛋，建議於崇光百貨購買台灣製造之魚圓較好吃。福建麵可於北角春秧街購買。

• When you cook the noodles, put them in cold water before you turn the heat on. The soup should be thick and flavourful for the authentic taste.

• Originated from Fuzhou, Fujian fish balls are stuffed with pork filling. I personally prefer Taiwan-made ones from Sogo Department Store because they taste better. You can get Fujian noodles from Chun Yeung Street Market at North Point.

福 建 • 甜 品
FUJIAN DESSERTS

與孫兒歡悦入廚，

滿室甜膩的香氣，

一份濃濃的溫馨情味！

Cooking with my grandchildren

filling the house with a syrupy fragrance

the close bonding and caring love

that is never too sweet

上元湯圓

Tangyuan
~ Glutinous Rice Balls
with Peanut
and Sesame Filling~

花生粉 4 杯
芝麻粉 4 杯
乾葱頭 1 1/2 杯（切片）
糖 1 1/2 杯（可加減）
豬油半杯
糖冬瓜 1 杯（切碎）
糯米粉 1 斤（1 包）

【 Ingredients 】
4 cups ground peanuts
4 cups ground sesames
1 1/2 cups sliced shallot
1 1/2 cups sugar
(adjust according to your taste)
1/2 cup lard
1 cup diced candied winter melon
600 g glutinous rice flour

【小秘訣 • TIPS】

- 原粒花生及芝麻的份量自行調整，
  攪拌後得花生粉及芝麻粉各 4 杯即
  可。

- 建議餡料的甜味濃一點，因外皮的
  糯米粉沒甜味。

- 芝麻花生球必須揘得緊實，滾糯米
  粉時才不會鬆散，建議將揘緊的花
  生球放入冰格待一晚，翌日才滾上
  粉。

- 芝麻花生球必須整個沾上水，才滾
  上糯米粉。

- 花生湯圓伴白開水吃，是泉州花生
  湯圓的吃法。

- Feel free to adjust the amount of whole
  peanuts and sesames as long as you get
  4 cups of ground peanuts and 4 cups of
  ground sesames at last.

- It's okay to make the filling a bit sweeter
  because the glutinous rice flour is bland
  and the glutinous rice balls are not served
  in syrup.

- The filling balls have to be squeezed
  compact. Otherwise, they may fall apart
  when you roll them in dry glutinous rice
  flour. It's best if you can refrigerate them
  overnight before rolling them in dry flour.

- Dip the whole filling balls into water before
  rolling them in dry glutinous rice flour.

- Tangyuan should be served in boiling
  water (not syrup). This is the authentic way
  to serve them in Quanzhou.

農曆十五是上元節，古時深居簡出的小姐在當天得父母允許上街看花燈，故多有才子佳人的美事，所以上元節也是中國情人節。吃一碗湯圓，祝願天下有情人甜蜜圓滿！

【 做法 】

1. 青島連殼脆花生去殼，連外膜放進預熱焗爐，用低溫 100℃ 焗 15 分鐘至香脆（試試是否香脆，建議待涼才試）待涼，去外膜。花生分成多份，放入攪拌機內攪成花生粉。

2. 芝麻傾入白鑊內，用小火炒至金黃，取出，攪成粉狀，與花生粉拌勻。

3. 燒熱鑊，下豬油半杯炒乾葱頭至金黃，將乾葱頭及油傾入花生芝麻粉內，再加入糖及糖冬瓜拌勻，試味。

4. 取芝麻花生餡料捏成球狀，若鬆散不好捏，可加點油拌勻帶黏性，芝麻花生球大小如焓熟的蛋黃，約可做 90 至 100 粒，冷藏。

5. 將糯米粉 2 至 3 杯放在深盤內（份量視乎弄多少湯圓），將芝麻花生球逐一放入水內，快速略浸一下即放入糯米粉內，雙手持盤轉動，讓花生球滾動沾滿糯米粉，取出，將花生球再浸入水內，快速取出，再次放入糯米粉內滾動，重複以上步驟共 4 次。燒滾水，放入花生湯圓，用大火煮滾至湯圓浮起，轉中火再煮 5 分鐘。

6. 上桌時，在大湯碗內傾入滾水，放入湯圓，滾水內不必加糖或任何調味，趁熱品嘗。

【 Method 】

1. To make ground peanuts from scratch, bake shelled Qingdao peanuts in a preheated oven at 100°C for 15 minutes until lightly toasted. (To test if the peanuts are crispy enough, wait till they are cold and taste them.) Leave them to cool and rub off the red coats. Divide the peanuts into small batches. Blend each batch in a blender at one time until fine. Set aside.

2. To make the ground sesame from scratch, stir fry sesames in a pan over low heat until golden. Leave them to cool. Blend until fine. Mix the ground sesame with ground peanuts.

3. Heat a wok. Pour in 1/2 cup of lard. Stir fry the shallot until golden. Pour the fried shallot and oil into the mixture of ground peanuts and sesames. Add sugar and candied winter melon. Mix well. Taste it to see if it needs more sugar. This is the filling.

4. Roll the filling into small balls. If the filling is crumbly and doesn't bind itself well, add a little oil and mix well. Each filling ball should be about the size of a hard-boiled egg yolk. This recipe should make 90 to 100 filling balls. Refrigerate until firm.

5. Put 2 to 3 cups of glutinous rice flour into a deep tray (the amount depends on how many dumplings you're making). Dip the filling balls one by one into water quickly and roll them in the dry glutinous flour to coat well. Then swirl the tray to roll the balls in the glutinous rice flour repeatedly. Then quickly dip the flour coated balls into water again and swirl them in the glutinous rice flour once more. Repeat this step 4 times in total. Then boil some water and put in the glutinous rice balls. Boil over high heat until the glutinous rice balls floats. Turn to medium heat and cook for 5 more minutes.

6. Before serving, fill a large soup bowl with boiling water. Put the cooked glutinous rice balls into the water. You don't need to season the water with sugar or anything else. Serve hot.

紅粿壽龜

Red
Tortoise
Sweet
Dumplings

一般而言，紅粿壽龜用於壽宴、婚嫁、拜觀音、拜祭祖先、許願及酬謝神恩等祭典上，紅彤彤的，很好看、也好吃！餡料除用綠豆茸外，也可選用花生芝麻，作為甜食也可以斑蘭葉汁做成綠色小龜。

【皮料】（約製成 18 個）
糯米粉 5 杯
雲呢拿香油 2 茶匙
砂糖 6 湯匙
鹽半茶匙
水約 2 至 3 杯
紅色食用色素 1 1/2 茶匙

【餡料】
香蕉油 1 1/2 茶匙
綠豆茸 600 克（做法參考第 159 頁）
糖 2 杯
油 2 湯匙

【配料及工具】
糉葉數片
糯米粉適量（掃模用）
餅模數個

【 Dumpling skin (makes 18 dumplings) 】
5 cups glutinous rice flour
2 tsps vanilla essence
6 tbsps sugar
1/2 tsp salt
2 to 3 cups water
1 1/2 tsps red food colouring

【 Filling 】
1 1/2 tsps banana essence
600 g mung bean paste (please refer to p.159 for method)
2 cups sugar
2 tbsps oil

【 Tools and supplies 】
a few bamboo leaves (as base support for the dumplings)
glutinous rice flour (for flouring the moulds)
a few tortoise cake moulds

【 做法 】

1. 取糯米粉半杯用水搓成小糰，分成三小塊，放入滾水內煮至浮起，取出，與其餘糯米粉及皮料用水搓成柔滑粉糰，柔軟程度如耳珠。

2. 綠豆茸、香蕉油、糖及油搓勻，弄成 20 個小圓球（每個約 28 克）。

3. 餅模內灑上糯米粉，取粉糰 56 克，擀成餅皮（中間薄一些，尺吋比餅模大），鋪在餅模上，用手略按中間有花紋的位置，放上 28 克餡料，將四周的皮料蓋於餡料上，再放上一片已掃油的糭葉，倒扣紅粿於碟上。

4. 紅粿放於蒸籠用小火蒸 12 分鐘，取出，掃上油即可。

【 Method 】

1. Add some water to 1/2 cup of glutinous rice flour. Knead into dough. Divide into 3 flat pieces. Blanch in boiling water until they float. Drain. Put the blanched dough into the rest of the dumpling skin ingredients. Add water and knead into smooth soft dough.

2. Mix banana essence with the mung bean paste, sugar and oil. Roll into 20 small balls (about 28 g each).

3. Flour the mould with glutinous rice flour. Take 56 g of the dough and roll it out thin. It should be larger than the mould with the centre thinner than the rim. Press the dumpling skin into the mould with your fingers (especially at the centre where the pattern is). Put 28 g of filling on the dumpling skin. Spread evenly. Fold the loose ends of the dumpling skin inward to cover the filling completely. Seal the seams well. Top with a greased bamboo leaf. Then turn the dumpling out onto the leaf. Arrange on a plate.

4. Steam the dumplings in a steamer over low heat for 12 minutes. Brush oil over them. Serve.

【小秘訣 • TIPS】

- 糯米粉糰搓得較乾，容易從餅模中扣出。
- 用小火蒸紅粿，餅模上印出的花紋才清晰好看。
- 倒模時，將餅模在砧板邊左右輕敲數下，容易脫出。
- 餅模於上海街售賣製餅工具的店舖有售。

- The dough for the dumpling skin should be a bit drier than usual, so that you can turn the dumpling out of the mould more easily.

- Steam the dumplings over low heat, so that the fine patterns on them can be kept intact.

- When you unmould the dumplings, tap the mould on the rim of a chopping board a couple times to loosen it first.

- The cake moulds are available from baking supply stores on Shanghai street.

炸番薯棗

Deep Fried
Sweet Potato
Dumplings

【材料】
　番薯 2 個（約 600 克）
　開邊綠豆 220 克

【番薯皮配料】
　糯米粉 4 杯
　糖半杯（視乎番薯甜度而增減）
　菜油或豬油 1 湯匙
　水 2 1/2 杯（可酌量減少，視乎番
　薯之乾濕程度）
　鹽半茶匙

【綠豆茸調味料】
　油 2 湯匙
　糖 3/4 杯
　鹽 1/4 茶匙
　香蕉油半茶匙

【 Ingredients 】

2 sweet potatoes (about 600 g)

220 g mung beans (split and hulled)

【 Dumpling skin ingredients 】

4 cups glutinous rice flour

1/2 cup sugar (to be adjusted according to the sweetness of the sweet potatoes)

1 tbsp vegetable oil or lard

2 1/2 cups water (to be adjusted according to the water content in sweet potatoes)

1/2 tsp salt

【 Seasoning for filling 】

2 tbsps oil

3/4 cup sugar

1/4 tsp salt

1/2 tsp banana essence

【 Method 】

1. Peel and slice the sweet potatoes. Steam until done. Mash them while still hot. Put in the rest of the dumpling skin ingredients. Knead into dough.

2. To make the filling, rinse the mung beans. Soak them in water for 1 hour. Drain. Steam for 30 minutes and stir in the seasoning while still hot. Mix well and mash.

3. Take 2 tbsps of the dumpling skin dough. Roll it round and make a hole at the centre. Stuff it with 1 tbsp of the mung bean filling. Seal the seam and roll it round. Shape them like dates without breaking the skin.

4. Deep fry the dumplings in warm oil over low heat. When they turn light golden, turn the heat up to high and deep fry until golden. It takes about 10 minutes to deep fry them.

【 做法 】

1. 番薯去皮、切片,蒸熟,趁熱壓成茸,加入配料搓成粉糰。

2. 開邊綠豆洗淨,用水浸 1 小時,瀝乾水分,蒸 30 分鐘,趁熱加入調味料拌勻,壓成綠豆茸。

3. 取番薯茸 2 湯匙搓圓,中間捏成窩形,包入綠豆茸 1 湯匙,搓圓,再捏成棗子形狀(別溢出綠豆茸)。

4. 放入溫油內用小火慢炸,取出前轉大火炸至金黃色即可,整個過程約需 10 分鐘。

家傳滋味 福建菜

作者	Author
王陳茵茵	Valerie Ong

策劃/編輯	Project Editor
	Karen Kan

攝影	Photographer
	Imagine Union

美術統籌及設計	Art Direction & Design
	Ami

出版者 Publisher

Forms Kitchen

香港鰂魚涌英皇道1065號　Room 1305, Eastern Centre, 1065 King's Road,
東達中心1305室　Quarry Bay, Hong Kong.
電話　Tel: 2564 7511
傳真　Fax: 2565 5539
電郵　Email: info@wanlibk.com
網址　Web Site: http://www.formspub.com
　　　　http://www.facebook.com/formspub

發行者 Distributor

香港聯合書刊物流有限公司　SUP Publishing Logistics (HK) Ltd.
香港新界大埔汀麗路36號　3/F., C&C Building, 36 Ting Lai Road,
中華商務印刷大廈3字樓　Tai Po, N.T., Hong Kong
電話　Tel:　2150 2100
傳真　Fax:　2407 3062
電郵　Email: info@suplogistics.com.hk

承印者 Printer

合群(中國)印刷包裝有限公司　Powerful (China) Printing & Packing Co., Ltd.

出版日期 Publishing Date

二O一O年十二月第一次印刷　First print in December 2010
二O一六年四月第二次印刷　Second print in April 2016